AMERICAN TANKS OF WORLD WAR II
1941–45

AMERICAN TANKS OF WORLD WAR II
1941–45
TANKS · SELF-PROPELLED GUNS · HALF-TRACKS · AMPHIBIOUS AFVs

STEPHEN HART & RUSSELL HART

First published in 2023

Reprinted in 2024

Copyright © 2023 Amber Books Ltd

All rights reserved. No part of this publication may be reproduced, stored in a retrieval system, or transmitted in any form or by any means, electronic, mechanical, photocopying, recording, or otherwise, without prior written permission of the copyright holder.

Published by Amber Books Ltd
United House
London N7 9DP
United Kingdom
www.amberbooks.co.uk
Facebook: amberbooks
YouTube: amberbooksltd
Instagram: amberbooksltd
X(Twitter): @amberbooks

ISBN: 978-1-83886-289-3

Editor: Michael Spilling
Designer: Andrew Easton
Picture research: Terry Forshaw

Printed in China

Contents

Introduction	6
Light Tanks	8
Medium Tanks	30
Heavy Tanks	60
Self-propelled Guns	70
Amphibious Landing Vehicles	88
Half-tracks and Armoured Cars	98
Half-tracked Self-propelled Guns	114
INDEX	125
PICTURE CREDITS	128

Introduction

When World War II started in Europe in 1939, the neutral USA merely retained a small backward force of tanks and armoured fighting vehicles (AFVs); yet by 1944–45 the many thousands of effective American AFVs deployed made a key contribution to victory in Europe and the Pacific.

World War II in Europe and the Pacific developed into a titanic clash of mass-mobilized industrialized nations. A key dimension to the conflict's outcome was the belligerents' ability to design, develop, manufacture, and deploy effective armoured fighting vehicles (AFVs). These vehicles spanned seven categories: from first to third, respectively, light, medium and heavy tanks; fourth, tank destroyers, tracked self-propelled guns (SPGs) and self-propelled anti-aircraft guns (SPAAGs); fifth, amphibious landing vehicles; sixth, half-tracked armoured personnel carriers (APCs) and wheeled armoured cars; and finally half-tracked SPGs and SPAAGs.

Mass production

The USA's development of AFVs played a key role in determining Allied victory during 1945. At the war in Europe's September 1939 start, neutral America languished behind most combatants in the size and effectiveness of its AFV inventory.

During the war's first phase, up to America's December 1941 entry into the conflict, the US tentatively expanded and modernized its inventory. It developed and deployed new AFVs, including: the M3 Stuart light tank; the stop-gap M2 and M3

An M18 tank destroyer exits a pontoon bridge established across a wide river in Northwest Europe late in the war; all of its five-man crew are visible in this image.

INTRODUCTION

A US Marine Corps modified Sherman M4A3 'Zippo' flame-throwing tank shoots its lethal jet of flame against a Japanese strongpoint during the battle for Iwo Jima, March 1945.

Lee/Grant medium tanks; the LVT-1 amphibious vehicle; the M2 and M3 half-track APCs; the improvised half-tracked T12 75mm (2.95in) Howitzer Motor Carriage (HMC) SPG; and the M4 81mm half-track Mortar Carrier.

New series

Between 1942 and mid-1944, the middle phase of America's war, the US enhanced its capacity to mass produce relatively more effective AFVs, many of which were sold to Allies like the United Kingdom, the Free French and the USSR. These somewhat more effective AFVs included: The M22 Locust airborne light tank; the M6 heavy tank; the M10 and M18 tank destroyers; the LVT-3 and LVT-4 amphibians; and the M13, M14 and M15 half-tracked SPAAGs.

Mighty Sherman

Furthermore, the key development during this phase of the war was that of the M4 Sherman medium tank, America's first genuinely well-balanced multi-role combat design. With an appropriate balance of adequate lethality, survivability and mobility, the M4 was well suited for mass production: During March 1942–August 1945 American firms manufactured some 49,300 Sherman tanks across multiple sub-variants.

The final phase of America's wars in Europe and the Pacific lasted from mid-1944 through to August 1945. In this phase, modest numbers of highly effective American AFVs began to reach the frontline.

These late-comers to the fight included the M24 Chaffee light and M26 Pershing heavy tanks. Other efficient AFVs included: the M36 tank destroyer; the LVT(A)-4 amphibious vehicle; half-tracked SPGs like the 105mm HMC M37 and the 155mm GMC M40; and the half-tracked M19 SPAAG.

America's ability to develop and deploy many thousands of reasonably effective AFVs across all these seven categories, as well as supply these AFVs to her Allies, played a crucial role in the Allied victories achieved in Europe and the Pacific by 1945.

LIGHT TANKS

LIGHT TANKS

Leading up to the start of World War II, the US Army possessed few operational light tanks. Despite this adverse situation, during 1939–43, American firms developed the reasonably effective M2 and M3/M5 Stuart designs. Moreover, during 1943–45, the M22 Locust and M24 Chaffee designs also entered American service, joining combat in both the European and the Pacific theatres.

The following tanks are featured in this chapter:
- Marmon-Herrington CTLM/CTLS
- T1 Cunningham
- M2 Light Tank
- M3 Stuart
- M3A1 Satan
- M5 Stuart
- M22 Locust
- M24 Chaffee

A US Marine Corps (USMC) M2A4 light tank operating with infantry on Guadalcanal island during late 1942; note the central- and rear-positioned hull superstructure atop which is a forward-located angular turret.

LIGHT TANKS

Marmon-Herrington CTLM/CTLS

The American company Marmon-Harrington produced 20 CTL-6 light tanks during 1938–39 for the US Marine Corps (USMC).

Next, during 1940–41 after the German occupation of the Netherlands, the Royal Dutch East Indies Army (RNIL) ordered 628 turreted CTL-6 variants designated CTLM/CTLS. During 1941 the US Army also ordered 240 Lend-Lease CTLS for the Chinese. Weighing 7.2 tonnes (7.1 tons), the CTLS included a low flat superstructure on top of tall tracks.

On the forward hull sat an angular turret that typically mounted a 7.62mm (0.3in) Browning M1919A4 machine gun. The design could also sport two additional ball-mounted machine guns in the front hull plus another one on the turret roof, meaning the design could sport between one and four Brownings. The tank featured light protection of just 12.7mm (0.5in) bolted plates.

Running gear

The tank's running gear comprised two wide-spaced Vertical Volute Spring Suspension (VVSS) bogies per side (each with two small road wheels), a high-positioned front driving wheel, a rear idler and one central return roller. The CTLS was powered by a rear-located 95kW (124hp) Hercules WXLC-3 petrol (gasoline) engine, enabling it to reach a top speed of 50km/h (31mph).

During early 1942, some 40 CTLS saw combat with RNIL during the Japanese conquest of the Dutch East Indies. A batch of 149 CTLS en route to RNIL at this time ended up serving in the Australian Army.

Marmon-Herrington CTLM/CTLS
Crew: 2
Production: 1938–41
Weight: 7.2 tonnes (7.1 tons)
Dimensions: (L,W,H): 3.5m (11ft 6in) x 2.08m (6ft10in) x 2.11m (6ft 11in)
Engine: 95kW (124hp) Hercules WXLC-3 petrol (gasoline)
Road speed: 50km/h (31mph)
Range: 161km (100 miles)
Armament: 1–4 x 7.62mm (0.3in) M1919A4 MG
Armour: 6.4–12.7mm (0.25–0.5in)

Marmon-Herrington CTLS
The Marmon-Herrington CTLS light tank had an unusual design feature; an untypically high superstructure floor, set above even the tops of its four small bogie-arrayed spoked road wheels.

T1 'Cunningham'

During 1927–31 the US Army contracted Cunningham to produce seven prototype T1 two-man light tank designs: the T1 and the T1E1-T1E6.

All seven shared the same basic compact hull design but with some modifications. The T1 had a rear-located superstructure on top of which sat the turret; in the E4 and E6 models, the latter features were positioned centrally. The front-engined E1-E3 and E5 models' running gear comprised a large rear driver-sprocket, four pairs of small bogie-suspended road wheels and a large-sized front idler. In the rear-engined and centrally-turreted E4 and E6 variants, the driver and idler wheel positions were reversed. The T1 and T1E1 were powered by a 78kW (105hp) Cunningham petrol (gasoline) engine and could achieve a top speed of 32km/h (20mph). All variants' turrets mounted a 37mm (1.46in) cannon alongside a co-axial 7.62mm (0.3in) Browning M1919A4 machine gun. The 7.6-tonne (7.5-ton) T1 featured 6.4–9.5mm (0.25–0.375in) thick armour.

T1 prototypes

The T1E1 featured an altered hull and relocated fuel tanks. The conical-turreted T1E2 featured 15mm (0.63in) thick armour. The T1E3 featured a redesigned suspension, while the centrally-turreted rear-engined T1E4 sported a new suspension based on four-wheeled leaf-spring bogies. The T1E5 incorporated the novel Cletrac differential steering system. Finally, the 10.1-tonne (9.95-ton) T1E6 had an up-rated engine and 16mm (0.63in) thick armour. The Army gained much valuable experience through using these T1 prototypes.

T1 'Cunningham'
Crew: 2
Production: 1927–31
Weight: 7.6 tonnes (7.5 tons)
Dimensions: (L,W,H): 3.8 (12ft 9in) x 3.8m (5ft 11in) x 2.1 (7ft 1in)
Engine: 78kW (105hp) Cunningham petrol (gasoline)
Road speed: 32km/h (20mph)
Range: 210km (130 miles)
Armament: 37mm (1.46in) M3; 7.62mm (0.3in) M1919A4 MG
Armour: 6.4–9.5mm (0.25–0.375in)

T1 'Cunningham'

This T1E1, seen in 1932, sports several distinctive features that differentiate it from its sister variants, including a large front idler wheel that extended beyond the superstructure's front and two prominent lamps on the hull front.

LIGHT TANKS

M2 Light Tank

The Light Tank M2 grew out of the T1 project of the late 1920s. During 1933 the Rock Island Arsenal (RIA) manufactured the first prototype T2 vehicle. This featured a Vickers-Horstman suspension with two sets of paired bogie wheels connected by leaf springs.

M2A4 Light Tank
A USMC M2A4 light tank in the drab olive camouflage scheme used during the 1942–43 Guadalcanal campaign. The vehicle's compact size is attested to by the size of the externally-mounted heavy machine gun.

RAI produced a second prototype vehicle, designated the T2A1, during early 1934. This vehicle utilized the new Vertical Volute Spring Suspension (VVSS) employed in the T5 experimental combat car (light tank), which had proven superior to the Vickers suspension in field trials. The T2A1 was powered by the compact Wright/Continental R670 aviation petrol (gasoline) engine.

The tank sported a fixed box-like 'turret' that mounted a 37mm (1.46in) cannon and a co-axial 7.62mm (0.3in) Browning M1919A4 machine gun; the hull front also carried a similar machine gun in a ball mount that was operated by the driver.

Variant modifications
During 1935, RAI produced 19 modified M2 tanks. These were essentially T2A1s with a different, smaller, one-man turret that merely sported as its main armament a single 7.62mm (0.3in) Browning M1919A4 machine gun; a similar weapon was fitted to the hull front. During 1936–37, American firms then manufactured 237 M2A2 variants. These vehicles were distinctive in that they sported twin rectangular-shaped side-by-side turrets, each mounting a single Browning M1919A4 machine gun. Next, some 73 10.2-tonne (10.1-ton) M2A3 variants were completed; these featured an improved VVS suspension

M2A4 Light Tank
Crew: 4
Production: 1938–39
Weight: 11.8 tonnes (11.6 tons)
Dimensions: (L,W,H) 4.43m (14ft 6in) x 2.47m (8ft 1in) x 2.64m (8ft 8in)
Engine: 186kW (250hp) up-rated Wright/Continental R670-9A petrol (gasoline)
Road speed: 58km/h (36mph)
Range: 322km (200 miles)
Armament: 37mm (1.46in) M5; 5 x 7.62mm (0.3in) M1919A4 MG
Armour: 6.4–25.4mm (0.25–1in)

LIGHT TANKS

with a wider gap between the bogie arrangements. Subsequently, during 1938–39, American firms produced several up-gunned and up-armoured prototype vehicles, which when accepted for mass production were designated the M2A4.

Power plant

Weighing 11.8 tonnes (11.6 tons), the M2A4 light tank had a crew of four: the vehicle commander/loader, the gunner, the driver and the co-driver. The vehicle was powered by a 186kW (250hp) up-rated Wright/Continental R670-9A seven-cylinder air-cooled radial petrol (gasoline) engine. Its running gear comprised relatively tall tracks fitted around two sets of paired VVSS-bogie medium-sized road wheels; a medium-sized close-spoked front driver; and a large open-spoked high-positioned rear idler. Rather than being horizontal, the top track rose slightly from the ends toward the middle, where it ran across two small return rollers. The tank's hull floor was also conspicuously high above the ground; this distance equated to four-fifths of its road wheels' height. The M2A4's power plant and suspension enabled it to obtain a top road speed of 58km/h (36mph). The tank could also achieve with a single load of fuel a maximum operational range of 322km (200 miles).

The M2A4 sported a relatively tall eight-sided welded and riveted angular turret. In the front of this was mounted a 37mm (1.46in) M5 cannon and a right-sided co-axial 7.62mm (0.3in) Browning M1919A4 machine gun. At the turret top rear stood a distinctive large angular commander's cupola. Most turrets were finished with an M20 bracket-mounting to support an external anti-aircraft Browning M1919A4 machine gun. In addition, to the right and left-hand sponson corners of the hull front superstructure the tank also featured another M1919A4 machine gun, as did the hull front. Taken together, this meant that the M2A4 light tank sported no fewer than five 7.62mm (0.3in) Browning machine guns; whether a four-man crew could effectively utilize this potential firepower remained moot. The

An M2 light tank being tested, circa 1940. On the side of the turret are marked a pair of white crossed swords, indicating that this tank belongs to the US Cavalry.

LIGHT TANKS

M2A1 Light Tank

The M2A1 variant sported a modified, rounded turret with a distinctive tall cylindrical commander's cupola located on the rear of the turret roof.

M2A1 Light Tank

Seen here in 1938, the M2A2 variant – dubbed by its crews the "Mae West" after the famous actress, featured two side-by-side turrets, one of which mounted a heavy machine gun and the other, a medium machine gun.

vehicle carried 8900 rounds of 7.62mm (0.3in) ammunition, plus 103 rounds for its 37mm (1.46in) M5 cannon. The tank sported 25mm (1in) thick frontal armour, but its sides and rear merely had 6.4mm (0.25in) thick plates.

American firms manufactured 375 M2A4s. During an 11-month production run that lasted from May 1940 through to March 1941, 365 vehicles were produced; during March 1942, a final batch of 10 tanks was completed. During 1941–42, Britain received 36 M2A4s but cancelled further deliveries in favour of the improved M3 Stuart, a natural development of the M2A4 that had by then served with the US Marine Corps (USMC) in the Pacific during 1942. These actions had revealed that the design was too under-armoured to be effective in combat conditions of the time. This realization led to the subsequent development of the M3 Stuart.

LIGHT TANKS

M3 Stuart

The American Light Tank – the M3 – was also known in British service as the Stuart after the famous Confederate Civil War General, J. E. B. Stuart; British units often referred to it as the Honey. September 1903.

The M3 grew naturally out of the aspiration to up-armour, up-gun and up-power the M2A4 light tank. After extensive trials of prototype vehicles over the winter of 1940–41, the American Car & Foundry Company (ACFC) completed the first main-production run vehicle during March 1941. In total, ACFC manufactured 5811 M3 tanks during a 15-month production cycle that lasted from March 1941 until June 1942. In British service they were officially designated as the Stuart Mk. I and Mk. II. The first batch of mass-production M3s featured levels of protection that were impressive for a light tank of 1941. These first M3s sported plates that were 44mm (1.7in) and 38mm (1.46in) thick on the lower and upper hull front, respectively. The turret armour was 51mm (2in) thick on the gun mantlet and 38mm (1.46in) thick on the sides. The hull sides and rear were also protected by 25mm (1in) thick armoured plates.

Armament

In terms of firepower, the first few hundred M3s manufactured carried the same weapons as the M2A4 tank. The turret sported the same 37mm (1.46in) M5 cannon, a co-axial 7.62mm (0.3in) Browning M1919A4 machine gun and a bracket-mounted external Browning machine gun. The two hull front sponsons each housed an additional

---

M3 Stuart Light
Crew: 4
Production: 1941–42
Weight: 12.7 tonnes (12.5 tons)
Dimensions: (L,W,H): 4.53m (14ft 10in) x 2.24m (7ft 4in) x 2.64m (8ft 8in)
Engine: 186kW (250hp) up-rated Wright/Continental W670-9A petrol (gasoline) or 191kW (250hp) Guiberson T-1020 diesel
Road speed: 58km/h (36mph)
Range: 193km (120 miles)
Armament: 37mm (1.46in) M5; 5 x 7.62mm (0.3in) M1919A4 MG
Armour: 9.5–44mm (0.375–1.7in)

M3 Stuart prototype

The prototype M3's new suspension and running gear distinguished it from its M2 predecessor; the rear large open-spoked idler wheel was positioned so it ran flat on the ground alongside the tank's four medium-sized road wheels.

LIGHT TANKS

M3A1 Stuart

This late-production M3A1 Stuart with shallow mud-guards, sports the tactical symbol "3", indicating it is from the 3rd USMC Tank Battalion; in this camouflage scheme it fought at Bougainville during November 1943.

Browning, while the fifth and final 7.62mm (0.3in) machine gun was ball-mounted in the tank's right-side hull front. Unlike the M2A4 design, the M5 cannon found in these early M3s benefited from an improved recoil system. From vehicle 314 onwards, the M3 instead carried the more potent longer-barrelled 37mm (1.46in) M6 L/56.6 cannon in an M40 mount. The M3 weighed 12.7 tonnes (12.5 tons) and was operated by a crew of four, namely the tank commander/loader, the gunner, the driver and the co-driver.

North Africa campaign

Some 4526 of the 5811 M3 light tanks manufactured were powered by the up-rated 186kW (250hp) Wright/Continental W670-9A seven-cylinder air-cooled aviation petrol (gasoline) engine. In contrast, the remaining 1285 M3s had the 191kW (250hp) Guiberson T-1020 four-stroke radial diesel engine as their power plant. These tanks were built to slightly modified British specifications and were dispatched under the Lend-Lease programme. In British service they were designated the Stuart Mk. II and fought in the North Africa campaign during 1941–42.

The early M3 was primarily distinguishable from the M2A4 by minor changes to its running gear. The basic track system was the same as the M2, namely two pairs of bogie VVSS road wheels, with a front driver and rear idler. Indeed, the M3 kept the M2's quirkiest feature – the front drive-sprocket wheel was positioned so close to the front-most road wheel that the two almost touched. Unlike with the M2, however, the M3's large, open seven-spoked rear idler was

---

M3A1 Stuart
Crew: 4
Production: 1942–43
Weight: 12.9 tonnes (12.7 tons)
Dimensions: (L,W,H): 4.53m (14ft 10in) x 2.24m (7ft 4in) x 2.39m (7ft 10in)
Engine: 186kW (250hp) up-rated Wright/Continental W670-9A petrol (gasoline) or 191kW (250hp) Guiberson T-1020 diesel
Road Speed: 39km/h (24mph)
Range: 217km (135 miles)
Armament: 37mm (1.46in) M5; 2 or 3 x 7.62mm (0.3in) M1919A4 MG
Armour: 9.5–51mm (0.375–2in)

positioned further to the rear and so low that it ran flat on the ground; it was also suspended on its own almost-horizontal leaf-spring arm. This combination of power plant, modified suspension and transmission enabled the early-production M3 to obtain a top speed of 39km/h (24mph) on roads.

Enhanced lethality

ACFC completed the first improved M3A1 light tank in May 1942. During a 10-month construction run that ended in February 1943, ACFC manufactured 4621 M3A1s at a rate of 462 vehicles per month. The four-crew M3A1 sported enhanced lethality modifications. These included an altered turret design that featured an internal basket in which the two turret crew sat, and rotated with the turret. The M3A1 also featured the 37mm (1.46in) M6 L/56.6 as standard but married to an improved gun vertical stabilizer. In addition, the two sponson-located hull front machine guns were dropped from the M3A1 as unnecessary extra weight.

M3A1s built for American service were powered by the 186kW (250hp) Wright/Continental W670-9A seven-cylinder air-cooled radial aviation petrol (gasoline) engine; these were known to the British as the Stuart Mk. III. However, M3A1 tanks manufactured in America by ACFC for British Service under Lend-Lease had the 191kW (250hp) Guiberson T-1020 diesel engine as their propulsion system; these were designated the Stuart Mk. IV in British parlance.

The M3A1 first saw sustained combat during the November 1942 Western Allied Torch landings. Tactical experience gained here showed that the M3A1's high-profile and non-sloped armour did not provide it with adequate survivability against the latest German armoured fighting vehicles (AFVs).

As an immediate stop-gap, between January–October 1943 American firms produced some 3427 examples of the modified M3A3 variant. This model – known to the British as the Stuart Mk. V – had a significantly redesigned hull that in places sported either better sloped or thickened armoured plates. These enabled weight savings through being thinner yet equally effective as thicker vertical plates.

Thus, the upper hull front went from 38mm (1.46in) sloped at 17 degrees to 25mm (1in) sloped at 48 degrees. Compared to the M3A1, these changes raised this variant's weight by 1.8 tonnes (1.77 tons) to 14.7 tonnes (14.5 tons). The hull redesign created more internal space, and this enabled the M3A3 to carry 174 rounds (rather than

M3A1 Stuart

This M3A1 Stuart fought in this olive drab camouflage scheme during the Western Allied Operation 'Torch' invasion of Vichy-controlled northwestern Africa; note the Stars-and-Stripes emblem painted on its hull side.

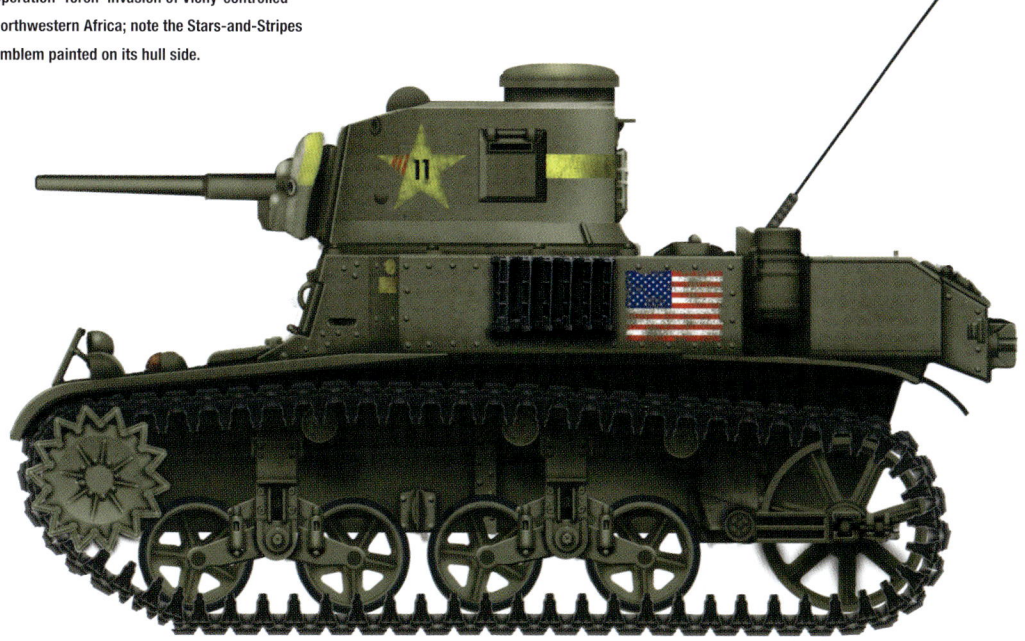

LIGHT TANKS

This photograph shows the crews of two USMC M3 Stuarts carrying out essential maintenance work – possibly munition re-stocking – on the edge of some dense jungle location.

106) for its 37mm (1.46in) cannon, as well as 7500 rounds (instead of 7220) of machine gun ammunition. These modifications extended the life of the M3 design as a scouting tank during 1943 but did little to alter the basic problem that the design was under-gunned and under-protected. In addition, shortages of petrol (gasoline) engines led to the simultaneous development of the diesel-powered Light Tank, M5 Stuart variant.

M3A3 Stuart
This M3A3 Stuart with redesigned hull is seen here during August 1944 sporting the red, white and blue tactical insignia of the First Free French Army.

M3A3 Stuart
Crew: 4
Production: 1943
Weight: 14.7 tonnes (14.5 tons)
Dimensions: (L,W,H): 4.53m (14ft 10in) x 2.24m (7ft 4in) x 2.39m (7ft 10in)
Engine: 186kW (250hp) up-rated Wright/Continental W670-9A petrol (gasoline) or 191kW (250hp) Guiberson T-1020 diesel
Road Speed: 58km/h (36mph)
Range: 217km (135 miles)
Armament: 37mm (1.46in) M5; 2 or 3 x 7.62mm (0.3in) M1919A4 MG
Armour: 9.5–51mm (0.375–2in)

M3A1 Satan

One unusual Stuart sub-variant was the M3A1 flame-throwing tank. During 1943, the USMC's combat experiences in the Pacific suggested that the M3 was becoming obsolescent and that flame weapons effectively rooted out determined well dug-in defending Japanese infantry.

USMC units first made improvised field modifications, mounting the M1A1 flame-thrower in either the space left by the removal of the bow front machine gun or through a turret pistol hole. Neither of these expedients proved particularly effective. Consequently, during late 1943 the US purchased 40 Canadian-made Ronson F.U.L. Mk. IV flame-throwing devices.

USMC mechanics installed a Ronson device in each of 24 surplus M3A1 tanks to create the Satan – so-named because of the terrifying flame it spewed forth. The short-barrelled Ronson device was mounted in the space left when the 37mm (1.46in) cannon was removed. The right-side co-axial machine gun was retained, while the bow machine gun was removed in some vehicles. Space inside the chassis was freed-up to enable the inclusion of a large 773 litre (170 gallon) fuel tank; the passage of the piping from this fuel reservoir to the Ronson device restricted the turret's traverse to just 180 degrees.

A squadron of 12 M3A1 Satans subsequently joined the USMC's 2nd and 4th Tank Battalions, alongside three new M5A1 light tanks for fire support tasks. These Satans were frequently employed in combat during the summer 1944 invasions of Saipan and Tinian.

M3A1 Satan
Crew: 4
Production: 1943
Weight: 12.9 tonnes (12.7 tons)
Dimensions: (L,W,H): 4.53m (14ft 10in) x 2.24m (7ft 4in) x 2.39m (7ft 10in)
Engine: 186kW (250hp) up-rated Wright/Continental R670 petrol (gasoline)
Road speed: 39km/h (24mph)
Range: 217km (135 miles)
Armament: Ronson F.U.L. Mk. IV flame-thrower; 1 x 7.62mm (0.3in) M1919A4 MG
Armour: 9.5–51mm (0.375–2in)

M3A1 Satan
Like many of the 24 M3A1 tanks converted into the Satan flame-throwing tank, this vehicle had its bow machine gun removed; unusually, however, it also sported a roof-mounted external M2HB heavy machine gun.

LIGHT TANKS

M5 Stuart

Between April and December 1942, Cadillac, General Motors and the Massey Harris Company manufactured a total of 2076 Light Tank, M5, Stuart vehicles between them.

M5 Stuart
M5 Stuart light tank in Tunisia during February 1943 during Operation 'Torch'; it is distinguished from its M3 predecessor by the raised rear decking, modified to accommodate the twin Cadillac engines and relocated fuel tanks.

Some 553 of these were dispatched to Allies under the Lend-Lease scheme, primarily to the British, but with some also going to the Free French forces. Weighing 15.1 tonnes (14.9 tons), the four-crew M5 Stuart featured a different propulsion system to its M3, M3A1 and M3A3 predecessors. The M5 was powered by twinned 110kW (148hp) Cadillac 44T24 eight-cylinder four-cycle petrol (gasoline) car engines. Each engine produced 331 joules (244ft-lbs) of net torque at 1200rpm. The main reason for this change of propulsion was the expected failure during 1942 of Wright/Continental's petrol (gasoline) engine production to meet the surging demand from booming AFV construction.

Hull redesign

To accommodate this new power plant, the vehicle's fuel tanks were relocated and reduced in size from 416 litres (110gal) in the M3A3 down to just 340 litres (89gal). The new engines also required a redesign of the hull rear deck, with this being raised by 40 per cent across the back third of the hull. This raised level of the rear decking was the most obvious external difference between the M5 and the three earlier M3 variants. These twinned Cadillac 44T24 engines provided the M5 Stuart with better mobility than the M3A3. The former could obtain a top road speed of 58km/h (36mph), some 8km/h (5mph) more than the M3A3. However, since the M5 had a reduced fuel capacity,

M5 Stuart
Crew: 4
Production: 1942
Weight: 15.1 tonnes (14.9 tons)
Dimensions: (L,W,H): 4.34m (14ft 3in) x 2.24m (7ft 4in) x 2.59m (8ft 6in)
Engine: Twinned 110kW (148hp) Cadillac 44T24 petrol (gasoline)
Road speed: 58m/h (36mph)
Range: 161km (100 miles)
Armament: 37mm (1.46in) M6; 3 x 7.62mm (0.3in) M1919A4 MG
Armour: 9.5–51mm (0.375–2in)

LIGHT TANKS

A 3rd US Armored Division M5 Stuart moves through the Vauterie hamlet of Saint-Fromond, Normandy, June 1944. In a surreal twist, nearby a child's doll has been placed on a stone next to the bullet-pocked wall of a house.

its maximum operational range on a single tank was merely 161km (100 miles), 35 per cent less than the 217km (135 mile) range of the M3A3.

The M5 sported the same firepower arrangements as the M3A3: a turret-mounted 37mm (1.46in) M6 cannon and a co-axial 7.62mm (0.3in) Browning M1919A4 machine gun; a tall M20 bracket-mounted turret-roof anti-aircraft Browning M1919A4 machine gun; and another 7.62mm (0.3in) Browning fixed in a M14 ball-mount located on the right side of the hull front. In addition, the level of protection sported by the M5 was almost identical to that of the M3A3: it featured turret armour that was 51mm (2in) thick on the gun mantlet and 32mm (1.25in) thick on the sides. Its lower hull front was protected by 44mm (1.75in) plates. The main difference in comparison to the M3A3 was that the M5's upper hull front was increased from 25mm (1in) thickness sloped at 48 degrees to 28.6mm (1.125in) at the same slant.

Firepower assets

During a 29-month production run that spanned from November 1942 to June 1944, the firms ACFC, Cadillac, General Motors and Massey Harris manufactured 6810 improved M5A1 light tanks at an average rate of 235 units per month. The 15.2-tonne (15-ton) M5A1 featured a redesigned elongated turret, with the M20 bracket mounting for the external Browning machine gun being moved to the right side of the turret roof rear. The turret side pistol port was discontinued in most M5A1s. In addition, the M5A1 carried more munitions than its M5 predecessor – this variant stowed some 146 rounds for its main gun and 6750 machine gun rounds. This aside, the M5A1 had the same firepower assets and its welded steel plates had the same arrangements as the M5.

The M5A1's running gear was also identical to its predecessor. However, the modestly increased weight did somewhat raise the tank's ground pressure rating. To address this, three M5A1E1 prototypes were

21

LIGHT TANKS

manufactured with various track widening improvisations but the results were not favourable and this programme was eventually cancelled. Other M5A1 modifications included the provision of hatches for the drivers that were equipped with a periscope.

The M5A1 provided useful service in the tactical roles of reconnaissance and scouting during 1943–45 in the Italian and Northwest Europe campaigns, as well as in the Pacific.

As the M24 Chaffee entered service in Europe in late 1944 and early 1945 it replaced many of the surviving M5 and M5A1 Stuarts then in front-line service.

An M5 Stuart of the US 12th Armoured Division in the town square of Rouffach, 5 February 1945, having overrun the German defence of the Colmar Pocket and linked up with Free French forces.

M5A1 Stuart

This early-production USMC M5A1 of the 4th Marine Tank Battalion, finished in an unusual three-tone camouflage scheme, fought on Roi-Namur island during the early February 1944 battle for Kwajalein Atoll.

LIGHT TANKS

M5A1 Stuart

Hailing from the US 501st Tank Destroyer Battalion's recce unit, this M5A1 fought during the Battle for the Volturno River, Italy, in October 1943; note the frontal mudguard and partial side-skirts.

M5A1 Stuart

This American M5A1, dubbed "Annie B" by its crew, prominently displays on its hull side the multi-coloured shield that is the tactical symbol of the 66th Armored Regiment.

LIGHT TANKS

M22 Locust

During 1941–42, the Americans developed an air-deployable light tank design in response to a British request for a replacement for their Tetrarch tank. A key criterion of the requirement was that the vehicle could be carried by the existing Douglas C-54 Skymaster transport aircraft or the British Hamilcar glider.

Initially, the firm of Walter Christie submitted an unusual design for a vehicle with well-sloped armour and four large road wheels married to the novel Christie suspension, but this was not accepted partly due to excessive weight.

Subsequently, during 1942, the firm of Marmon-Herrington developed a T9 prototype and two further T9E1 experimental vehicles. After trials the design was accepted into American service as the Light Tank (Airborne), M22. Ultimately, from an initial contract for 1600 vehicles, Marmon-Herrington manufactured some 830 M22 tanks during 1943–44, of which 230 were sent to the United Kingdom under the Lend-Lease programme. In British service it was designated as the M22 Locust.

Light protection

The M22 was a compact vehicle with tall tracks, and on top of these a shallow flat-topped superstructure. On this was set a forward-positioned cylindrical turret that featured an M54 mounting the ubiquitous 37mm (1.46in) M6 cannon as carried in the M3/M5 Stuart light and M3 Lee/Grant medium tanks. The tank carried 50 rounds for this gun. The turret also sported a right-positioned co-axial 7.62mm (0.3in) Browning M1919A4 machine gun, for which it stowed

Light Tank (Airborne) M22 Locust
Crew: 3
Production: 1943–44
Weight: 7.3 tonnes (7.2 tons)
Dimensions: (L,W,H): 3.94m (13ft) x 2.25m (7ft 12in) x 1.84m (6ft)
Engine: 121kW (162hp) Lycoming O-435T petrol (gasoline)
Road speed: 56m/h (35mph)
Range: 180km (110 miles)
Armament: 37mm (1.46in) M6; 1 x 7.62mm (0.3in) M1919A4 MG
Armour: 9.5–25mm (0.375–1in)

M22 Locust
Vehicle 3066998 "Bonnie" was one of a minority of American M22 Locust light tanks manufactured without side-skirts; it served with the US 28th Airborne Tank Battalion.

LIGHT TANKS

M22 Locust

This M22 Locust sports the shallow side-skirts – barely covering the top track – routinely fitted to the design during manufacture; the distinctive horizontal wheel brace is also evident here.

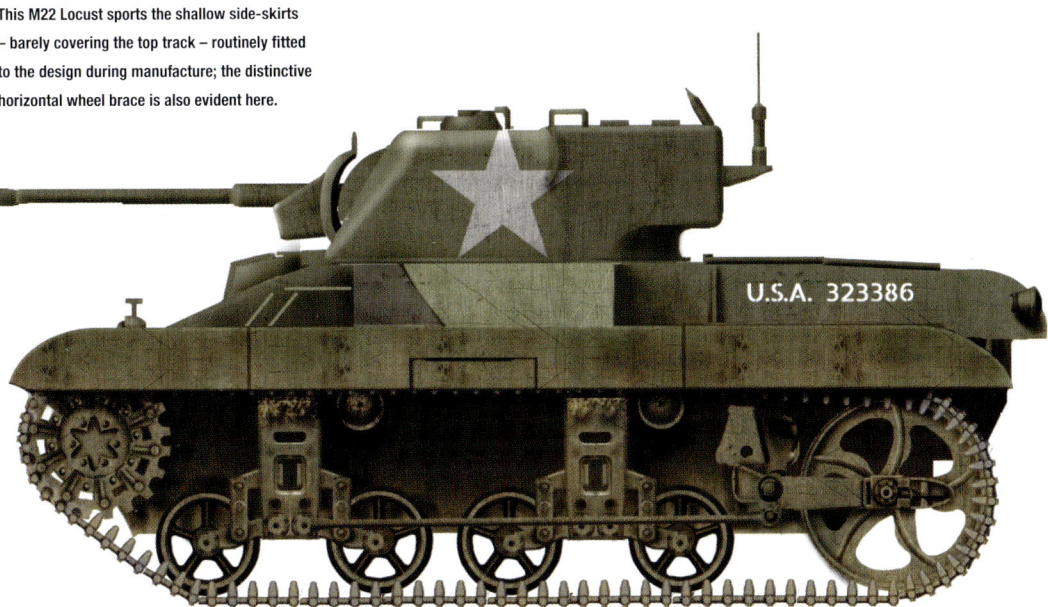

2500 rounds. Despite its compact dimensions, the M22 still weighed 7.3 tonnes (7.2 tons).

It was manned by a crew of three: the commander/loader was in the turret-right position, the gunner in the turret-left and the driver in the hull-left. The Locust featured light protection of just 9.5-25mm (0.375-1in) steel Rolled Homogenous Armour (RHA) plates.

Running gear

The Locust's running gear comprised two widely-spaced VVSS bogies per side (each with two small road wheels), a high-positioned 22-tooth small front driving wheel, a large rear idler, two small upper return rollers and an unusual final feature – a horizontal strengthening brace. The vehicle was powered by a 121kW (162hp) Lycoming O-435T six-cylinder petrol (gasoline) engine.

This power plant enabled it to reach a top speed of 56km/h (35mph).

Impractical design

Ultimately, the Locust turned out to be an unsatisfactory design. It took 24 minutes to load an M22 onto a Skymaster transport because the turret had to be removed. It also proved to be mechanically unreliable and its armour protection was woefully inadequate for combat in 1944.

Brief service

Just 22 American M22 tanks were deployed to Europe during 1944. In addition, eight glider-borne British Locust tanks participated in Operation Varsity, the March 1945 Allied airborne landings executed to support the amphibious assault of the Rhine River code-named Operation Plunder.

LIGHT TANKS

M24 Chaffee

During 1941–42, the American Army began tentative development of a new advanced light tank that would replace the M3/M5 Stuart then in production. Any new design needed to have improved firepower and survivability to be able to function effectively in the classic light tank doctrinal role: that of armoured reconnaissance and scouting.

A specification for a future 14-tonne (13.9-ton) low-silhouetted light tank that had armour of up to 38mm (1.46in) thickness and a 37mm (1.46in) cannon was issued during 1941. RIA produced the first five prototype vehicles during 1941–42 designated as the T7 and the T7E1-T7E4. These experimental vehicles featured different arrangements of hull/turret composition, suspensions and engines. Trials showed that the T7E2 variant had the most potential. This tank, powered by a 224kW (300hp) Wright/Continental R975 engine, sported a cast hull and turret. It was decided to up-gun the design to include the 57mm (2.2in) cannon due to be fitted in the Canadian Army's Ram tank. Subsequent variants of what became the M7 were further up-gunned and up-armoured, leading it to be re-classified as a medium tank. With the Sherman coming on-stream the M7 project was cancelled. However, the design work nevertheless assisted the development of what became the M24 light tank.

During 1943, the Ordnance Department worked with Cadillac to design a new light tank that incorporated the lessons learned from the T7/M7 developmental work. In October 1943 Cadillac produced the first prototype T24 light tank. Extensive trials showed that it had good potential and in December

M24 Chaffee
Crew: 5
Production: 1944–45
Weight: 18.3 tonnes (18 tons)
Dimensions: (L,W,H): 5.03m (16ft 6in) x 3m (9ft 10in) x 2.77m (9ft 1in)
Engine: twinned Cadillac 110kW (148hp) 44T24 eight-cylinder four-cycle petrol (gasoline)
Road speed: 56km/h (35mph)
Range: 161km (100 miles)
Armament: 75mm (2.95in) M6; 1 x 12.7mm (0.5in) M2HB HMG; 2 x 7.62mm (0.3in) M1919A4 MG
Armour: 9.5–25mm (0.375–1in)

M24 Chaffee
An American M24 Chaffee of the 740th Tank Battalion's Cavalry Reconnaissance Squadron in distressed white over olive camouflage during the December 1944 Battle of the Bulge.

LIGHT TANKS

An M24 Chaffee of the US 9th Armored Division, Germany, 1945. The tank, heavily laden with external bundles, has seven American personnel assembled around it or on it.

the Army placed an order for 1000 vehicles with Cadillac and Massey-Harris. Between March 1944 and September 1945 these firms manufactured 4731 M24s. The vehicle was subsequently designated the Chaffee after Major-General Adna Chaffee, one of the key guiding forces behind the development of the US armoured force.

Production variant

The M24 production variant featured a five-strong crew, consisting of commander, gunner, loader, driver and radio operator/co-driver. The tank sported fairly tall tracks, a shallow hull superstructure above them and a large conical-rounded turret that extended for two-thirds of the vehicle's length; the turret was topped with a large somewhat conical cylindrical commander's cupola. In terms of its physical dimensions, the 2.77m (9ft 1in) tall tank was 5.54m (18ft 2in) long and 2.98m (9ft 8in) wide. The M24 weighed 18.3 tonnes (18 tons), just over the stipulated weight for the design. To avoid being any heavier, the tank sported armoured plates that were merely 25mm (1in) thick almost everywhere.

The comparatively large turret featured in an M64 mounted the lightweight 75mm (2.95in) M6 L/39 gun; the cannon's concentric recoil saved space in the turret. Adjacent to this weapon was a co-axial 7.62mm (0.3in) Browning M1919A4 machine gun. In terms of other firepower, the M24 sported a second 7.62mm (0.3in) Browning machine gun in a ball mount located in the lower right hull. Finally, the vehicle sported a bracket-mounted 12.7mm (0.5in) Browning

LIGHT TANKS

M2HB heavy machine gun on the rear of the turret's roof.

Running gear

The tank's running gear comprised five large torsion-bar suspended road wheels, a high-positioned medium-sized front driving-wheel, a large rear idler and three small upper return rollers. Powered by twinned Cadillac 110kW (148hp) 44T24 eight-cylinder four-cycle petrol (gasoline) engines, the Chaffee could obtain a maximum on-road speed of 56km/h (35mph); when moving across country it could reach an impressive top speed of 40 km/h (25mph). With a full fuel load the tank could also achieve a maximum operational range of 161km (100 miles).

M24 tanks first saw combat with American forces in the December 1944 Battle of the Bulge. In American service they often served in the Cavalry Reconnaissance Squadrons found within armoured divisions. The Americans also dispatched 302 Chaffees to Britain (where they replaced M3 Stuarts), and also after the war's end a further 52 to the recently reestablished French Army. The quality of the M24's design is attested to by the fact that it remained in NATO service into the 1960s.

A group of five M24 Chaffee tanks has been quite precisely lined up with hatches open in winter 1944; the tanks' five-man crews appear busy checking over their vehicles.

LIGHT TANKS

M24 Chaffee

This M24 of the US 1st Armored Division in Bologna, Italy, late April 1945 sports the alternate pattern side-skirts, which are deeper (and sloping upwards) across the front and rear sections.

M24 Chaffee

The compact size of the Chaffee is illustrated in this unidentified M24 seen in Germany during early 1945; the length of its hull is merely double that of the external Browning M2HB heavy machine gun mounted on the turret.

MEDIUM TANKS

MEDIUM TANKS

From humble beginnings with the M2 medium tank, the Americans went on to develop the useful stop-gap M3 Lee/Grant. Next, from 1943 US firms developed and mass-produced the M4 Sherman. America's first genuine dual purpose battle tank, the tens of thousands of Sherman variants produced made a significant contribution to the eventual Western Allied victory during 1945.

The following tanks are featured in this chapter:
- M2
- M3 Lee/Grant
- M3 Lee Canal Defence Light (CDL)
- M4 Sherman
- M4A3R5 Flame-thrower Tank
- USMC Improvised M4A2 Flail Tank
- M4A3E2 'Jumbo' Assault Tank
- M7
- T23

German civilians watch M4A3 Sherman tanks supporting the 70th Infantry Division (from Seventh United States Army) drive down Dachauer Strasse, Munich, Germany, following the city's occupation on 30 April 1945.

MEDIUM TANKS

M2

The September 1939 beginning of World War II prompted the US Army to contract RIA (Rock Island Arsenal) to produce the M2 medium tank, a logical development of the M2 light tank. The high-silhouetted M2 featured a tall frontally-sloped angular box-like hull superstructure with a small turret on top.

This mounted a 37mm (1.46in) M3 cannon, which could penetrate 46mm (1.8in) of armour at 460m (500yds). The six-crewed M2 tank was distinctive because it carried between seven and nine machine guns: four sponson-mounted weapons in the upper hull front and rear corners, two driver-fired glacis-plate mounted ones, a co-axial turret weapon and two bracket-mountings that allowed for further weapons to be fixed on the hull roof. The tank also featured one other bizarre feature: metal deflector plates over the rear fenders. The idea was that when the tank drove over an enemy trench, the rear-facing sponson machine guns could fire and the rounds deflected down into the trench; the notion proved tactically useless.

Its running gear consisted of a frontal drive wheel, a rear idler and three external bogies with a VVSS arrangement each serving a pair of small road wheels. Powered by a 254kW (340hp) Wright/Continental R975-EC2 air-cooled radial petrol (gasoline) engine, the M2 could obtain a top speed of 43km/h (26mph). It carried 473 litres (104gal) of fuel, which gave the vehicle a maximum operating range of 210km (130 miles). Army trials conducted with the first batch of 18 M2s produced by RAI led to the realization that modifications were needed. Combat in Europe also suggested that the design was likely to be of tactically limited value just as it entered service, as it was inadequately armed, armoured and powered.

M2 Medium Tank
Crew: 6
Production: 1940
Weight: 17.25 tonnes (17 tons)
Dimensions: (L,W,H): 5.38m (17ft 8in) x 2.62m (8ft 7in) x 2.84m (9ft 4in)
Engine: 254kW (340hp) Wright/Continental R975-EC2 petrol (gasoline)
Road speed: 43km/h (26mph)
Range: 210km (130 miles)
Armament: 37mm (1.46in) M3; 7–9 x 7.62mm (0.3in) M1919A4 MG
Armour: 6.4–32mm (0.25–1.2in)

M2 (early model)
The additional external turret-side mount for an M1919A4 machine gun identifies this as an early-production M2; this variant possessed nine such weapons – more than its four gunners could ever effectively operate.

MEDIUM TANKS

M2A1

The improved M2A1 variant, seen here in early 1941, was easily distinguished from its predecessor by the new turret it sported as well as by some minor gun mantlet adjustments.

Up-powered variant

Between October 1940 and August 1941 RAI manufactured a further 94 examples of a modified version, the M2A1. This was an up-powered and up-armoured variant of the 18 M2s already in service. The M2A1 sported the larger turret of the M3 light tank and had gun mantlet armour that was increased to 51mm (2in) thickness. This variant also featured a turbo-charged 260kW (350hp) Wright/Continental R975-C1 engine. These modifications did little to address the key problem that the M2's main armament was inadequate.

With the more effective M3 Lee medium tank coming into production the US Army decided merely to use the 112 M2/M2A1s in existence as training vehicles within the United States. Although these tanks were only used for training purposes, the design nonetheless tested well and proved several key design features. These were incorporated into subsequent designs, notably the M3 Lee/Grant and the M4 Sherman.

This photograph of an early-production M2 tank nicely depicts the four half-cylindrical sponson mounts, each for a 7.62mm (0.3in) Browning machine gun, located in the front and rear corners of the hull superstructure.

MEDIUM TANKS

M3 Lee/Grant

The American Army's increasing doubts regarding the inadequacy of the firepower and protection capabilities of the tall-silhouetted M2 led to the development of the Medium Tank, M3. The US Army's variant of this tank (sporting the American-pattern turret) became known as the M3 Lee, named after the Confederate US Civil War commander, General Robert E. Lee.

M3 Grant Mk. I

This M3 Grant Mk. I of the British Eighth Army is seen here in October 1942 during the Gazala battles, North Africa. It sports a 'disruptive' camouflage scheme of khaki and mid-brown separated by a thin white outline.

The British Army's M3 variants sported the British-pattern turret and were dubbed the Grant, after the famous Union General Ulysses S. Grant. American firms built the M3 for the US Army and to a modified specification for the British under Lend-Lease. Between June 1941 and February 1942, the firms of American Locomotive (ALCO), Chrysler's Detroit Tank Arsenal and RAI constructed some 3512 vehicles of the 27.9-tonne (27.5-ton) M3 Lee design. Simultaneously, Baldwin, Pressed Steel Car (PSC) and Pullman manufactured 1212 British Army 28.1-tonne (27.7-ton) M3 Grant Mk. I tanks. Total M3 production was therefore some 4724 tanks. Four other sub-contractor firms manufactured M3 turrets: each firm's turrets had minor differences and were built to either the American pattern (M3 Lee) or the British pattern (M3 Grant).

Development commenced in July 1940 using the M2 design as the starting point. Such was the Allied demand for medium tanks that the whole process was rushed to get a stop-gap vehicle into the front line as quickly as possible. The Army had requested that the new medium tank in production include a potent 75mm (2.95in) gun as its main armament. American defence firms, however, lacked experience in turret design with such a large-calibre weapon. The solution was to mount the new

M3 Mk. I Grant
Crew: 7
Production: 1941–42
Weight: 27.9 tonnes (27.5 tons)
Dimensions: (L,W,H): 5.64m (18ft 6in) x 2.72m (8ft 11in) x 3.12m (10ft 3in)
Engine: 254kW (340 hp) Wright/Continental R975-EC2 petrol (gasoline)
Road speed: 34km/h (21mph)
Range: 193km (120 miles)
Armament: 75mm (2.95in) M2; 37mm (1.46in) M5 or M6; 4–5 x 7.62mm (0.3in) M1919A4 MG
Armour: 13–51mm (0.5–2in)

prototype tank's 75mm (2.95in) cannon in a limited-traverse mounting in an offset sponson located in the right front of the hull. On top of this hull was mounted a typical American modestly-sized turret that housed the 37mm (1.46in) M5 cannon. Set on top of the turret was a small cupola that housed a 7.62mm (0.3in) Browning M1919A4 machine gun. This unusual arrangement of a hull-mounted main armament was not unknown in the conflict's early phases, as seen, for example, in the French Char 1 tank.

High silhouette

The M3 featured a tall, riveted hull superstructure on top of which sat the turret and on top of that the machine-gun cupola, giving the vehicle a high profile. The tank featured a well-sloped (45-degree) lower hull glacis plate and then a slightly-sloped (37-degree) upper hull front, in which to the right sat the large vertical-faced sponson that housed the limited traverse 75mm (2.95in) gun. The horizontal hull roof stretched for about 40 per cent of the vehicle's length. Located on the rear two-thirds of the hull roof was the distinctive fully-rotating M3 turret with its 37mm (1.46in) cannon. On top of this was the drum-like machine-gun cupola. These elements together gave the design a high silhouette coming in at 3.12m (10ft 3in) in height. The turret had a steeply-sloping front face and rounded edges. At the rear the hull superstructure took a shallow step downwards with the hull roof then sloping gently downwards to the back of the vehicle.

The sponson sported a limited-traverse 75mm (2.95in) M2 gun in the M1 mount; the gun could merely be traversed 15 degrees left or right from its centre-line. The weapon could also be manually elevated from -9 to +20 degrees. The 37mm (1.46in) M5 cannon, affixed in an M24 mount to the rotating turret's front, could be manually elevated from -7 to +60 degrees. The vehicle carried 46 rounds for its main gun, 178 shells for its 37mm (1.46in) secondary armament and 9200 machine-gun bullets. In addition, the design typically featured a minimum of four 7.62mm (0.3in) Browning M1919A4 machine guns: one was set in the commander's cupola on top of the rotating turret, one was co-axial to the 37mm (1.46in) cannon and a pair were located in the hull front.

In addition, some M3s were fitted with an external mounting for an anti-aircraft Browning M1919A4 machine gun at the turret rear. The hull of the M3 was protected by 13–51mm (0.51-2in) welded steel Rolled Homogenous Armour (RHA) plates. In addition, the turret featured cast steel RHA that

M3A5 Lee

This American M3A5, the last variant produced, is set here in Burma in a distressed field-grey camouflage scheme; the vehicle sports smoke mortar launchers fitted to its turret sides.

MEDIUM TANKS

M3 Lee (early production)
Seen here during Operation 'Torch' near Souk El-Aaba, Tunisia in November 1942, this early production M3 Lee, was fielded by the 13th Armored Regiment of US 1st Armored Division.

was 22–51mm (0.875–2in) thick. The British-pattern (Grant) turrets had a several minor modifications. They featured a simple hatch instead of the commander's cupola, a distinctive rear bustle that housed a Wireless Set No.19 and a side-mounted 51mm (2in) smoke mortar.

Design features

In terms of its running gear, the seven-crewed M3 Lee borrowed many of its components from its M2 predecessor. The M3 sported three pairs of small-sized open-spoked VVSS bogie road wheels on each side; the top of the track ran along three small return rollers that were affixed to the top of the bogie structure rather than to the tank hull itself. The M3 also featured a closed-spoked 13-tooth front-drive sprocket and a medium-sized open-spoked rear idler. The M3 was powered by a rear-located 254kW (340hp) Wright/Continental R975-EC2 nine-cylinder air-cooled radial petrol (gasoline) engine. This power plant and suspension enabled the M3 to obtain a maximum sustained road speed of 34km/h (21mph). On a single tank of fuel the vehicle had a top operational range of 193km (120 miles). In late production M3s, the two hull-mounted machine guns were deleted. Finally, the running gear of very late production M3s featured strengthened and elongated Vertical Volute Springs (VVS), as well as return rollers positioned at the top rear (not top centre) of the bogie structure.

Starting from January 1942, new variants of the M3 Lee/Grant began to be manufactured. First, during January–May 1942 American firms manufactured 302 improved M3A1 Lee/Grant tanks. The M3A1 (also known to the British as the Lee Mk. II) sported a cast rounded-edged hull that had thicker armour in some places. The fully-rotating turret was also redesigned with a slightly lower profile. Meanwhile, Baldwin and RAI also produced just 12 examples of the M3A2 (Lee Mk. III) variant during January 1942. This 27.4-tonne (27-ton) variant had a sharp-angled welded hull intended to simplify the manufacturing process. Subsequently, during March–May 1942 American firms also produced 323 M3A3 tanks known as the Lee Mk. V in British parlance. These were diesel-engined variants that featured a welded hull. Their power plant was a General Motors 6046 diesel engine that paired two 127kW (170hp) GM 6-71 six-cylinder engines into a common

MEDIUM TANKS

M3 Lee

One of the quirkier features of the M3 design was the fully-rotating cupola mounted on top of the turret roof that housed a 7.62mm (0.3in) Browning M1919A4 machine gun, seen here on 'Kentucky' in Tunisia, December 1942.

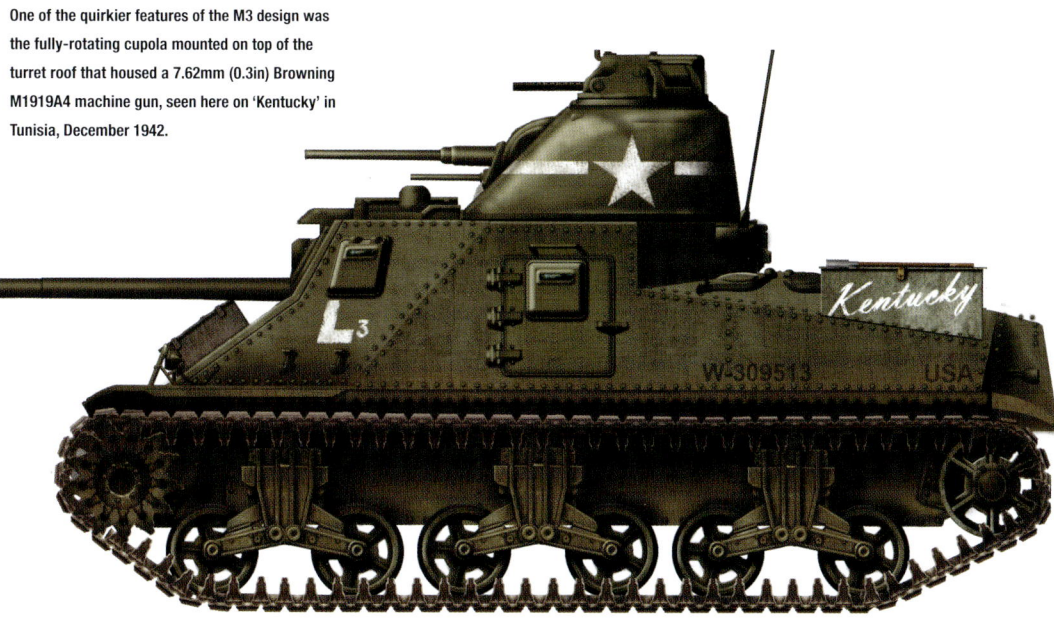

M3 Lee

M3 Lee of F Company, 12th Battalion, 3rd Regiment, US 1st Armored Division is seen here in Tunisia in February 1943 sporting a rare angular zig-zag disruptive camouflage scheme; note the compensating weight (not a muzzle-brake) at the end of the shorter-barrelled version of its 75mm (2.95in) main gun.

MEDIUM TANKS

unit. The British had also listed the designation of the Lee Mk. IV as an M3A3 variant that was powered by a Wright/Continental petrol (gasoline) engine; although planned, none of these Lee Mk. IVs were completed.

M3A5 variant

Finally, during January–May 1942, Baldwin constructed some 591 examples of the 29-tonne (28.5-ton) M3A5 variant. Most of these vehicles served in the British Army under the designation the Grant Mk. II. These variants featured a welded hull, the American-pattern turret and diesel propulsion. This variant was also powered by a General Motors 6046 diesel engine, namely paired 127kW (170hp) GM 6-71 six-cylinder engines. The shortage of Wright/Continental petrol (gasoline) engines largely explained this change in propulsion.

The M3A5 had a slightly enhanced mobility – it could obtain a top sustained road speed of 48km/h (30mph) and a maximum operating range of 241km (150 miles). In total, America sold 2855 M3 Lee/Grants to Britain and a further 1396 vehicles to the Soviet Union.

Although the M3 Lee/Grant was always seen as a stop-gap design until more effective multi-purpose medium tanks were developed, it nevertheless provided sterling service. It was the mainstay of the British Eighth Army in North Africa during 1942, spearheading key battles such as Gazala and El Alamein. M3s fought with American units in Tunisia during late 1942, most notably at Kasserine Pass and in the Pacific (for example, in the Gilbert Islands battles). During 1943, the M3 was phased out of front-line service as the M4 Sherman came on stream in large numbers. By March 1944 the American Army had declared the M3 obsolete. The British Army meanwhile had transferred 1706 M3 Grants to the Burma theatre, where they served until the war's end in the summer 1945.

M3 Lee
Crew: 7
Production: 1942
Weight: 29 tonnes (28.5 tons)
Dimensions: (L,W,H): 6.15m (20ft 2in) x 2.64m (8ft 8in) x 3.12m (10ft 3in)
Engine: 254kW (340 hp) Wright/Continental R975-EC2 petrol (gasoline)
Road speed: 34km/h (21 mph)
Range: 193km (120 miles)
Armament: 75mm (2.95in) M2; 37mm (1.46in) M6; 4–5 x 7.62mm (0.3in) M1919A4 MG
Armour: 12.7–51mm (0.5–2in)

M3 Lee

This image of an M3 during Operation 'Torch' during December 1942 shows the design's tall silhouette; this rendered the vehicle more vulnerable to detection and enemy anti-tank fire.

M3 Lee Canal Defence Light (CDL)

During 1940–42 the British developed a novel armoured warfare dimension – tanks that shone powerful lights intended to temporarily blind the enemy. This top-secret project was disguised under the innocuous-sounding cover name of Canal Defence Light (CDL) I.

M3 Lee (CDL)
This rare top secret M3 Lee variant is easily identified by the tall cylindrical turret that mounted its 'secret weapon' – a very powerful searchlight; the design, however, never 'blinded' the enemy in actual combat.

In American service the same vehicle was designated as the T10 Shop Tractor. Back in 1937 the concept of an AFV that 'fired' dazzling light had been demonstrated. However, it was not until 1940 that Britain developed a CDL variant of the Matilda Mk. II tank, whose rotating turret mounted a powerful searchlight. As 304 Matilda CDL vehicles were constructed, simultaneously design work unfolded during 1942 on a CDL variant of the M3 Lee/Grant tank.

In terms of external appearance, the chassis of the four-crew M3 Lee CDL tank was identical to standard M3/M3A1 tanks; British Grant CDL tanks tended to use the former chassis, while the Americans generally employed the latter. However, on top of the left front hull roof was a tall cylindrical CDL turret with a rounded right-hand side and a flat left-hand side. This turret looked very different from the standard small sloping-backwards turret of the M3 combat tank that mounted its 37mm (1.46in) cannon.

In the centre of the CDL turret front was a vertical search-light aperture. To compensate for this visual difference, British-manufactured CDL turrets sported a dummy 37mm (1.46in) cannon to the aperture's right. British M3 Grant CDLs also sported a

**M3A1 Lee Canal Defence Light/
T10 Shop Tractor**
Crew: 4
Production: 1942
Weight: 27.8 tonnes (27.4 tons)
Dimensions: (L,W,H): 6.15m (20ft 2in) x 2.64m (8ft 8in) x 3.24m (10ft 8in)
Engine: 254kW (340hp) Wright/Continental R975-EC2 petrol (gasoline)
Road speed: 34km/h (21mph)
Range: 193km (120 miles)
Armament: 75mm (2.95in) M2; 13 million candle-power (163.36 million Lumen) search-light; 1 x 7.62mm (0.3in) M1919A4 MG
Armour: 12.7–51mm (0.5–2in)

MEDIUM TANKS

block-mounted 7.9mm (0.31in) BESA machine gun to the aperture's left. Instead of the BESA, American M3 CDLs featured a ball-mounted 7.62mm (0.3in) Browning M1919A4 machine gun; many examples also featured no fake barrel. In addition, the M3 CDL retained the hull right-side sponson-mounted hull 75mm (2.95in) gun, and thus it retained its usual lethality that was augmented by the illumination weapon. The M3's CDL turret housed a 13-million candle-power (163.36 million lumen) carbon-arc searchlight and the solitary turret crewman. In the centre of the turret front, slightly offset to the right, there was a 610mm (24in) tall by 51mm (2in) wide aperture, through which the light was focused. A shutter automatically opened and closed twice a second, thus creating a flickering effect that dazzled any observer. Blue or amber filters could be rapidly cycled, further enhancing the dazzling effect.

Beyond the aperture, the beam of light diverged horizontally at 19 degrees and at 1.9 degrees; this meant that at 910m (2985ft) the vehicle projected a rectangle of blinding light that covered an area 31m by 311m (102ft by 1020ft).

An entire brigade of Matilda, Churchill, M3 Lee and M4 Sherman CDL tanks deployed to Normandy during 1944, but these strange vehicles were never ever employed in combat in this bespoke specialized role.

Two Sherman tanks exit through the massive front double doors of the beached Landing Ship Tank (LST) US-77 during the Allied operations in the Anzio bridgehead, Italy, early 1944. Note how the crews have stowed spare track sections on the glacis plate, in part to augment the tanks' levels of frontal protection.

MEDIUM TANKS

M4 Sherman

The M4 was designed as the successor to the M3 Lee/Grant medium tank. Developmental work on the M4, which was initially styled the T6, started in March 1941. The prototype T6 used the existing M3 chassis but with a new superstructure, and on top of this a turret that sported a low-velocity 75mm (2.95in) M3 gun.

The prototype trialled in September 1941 and the next month was standardized as the M4 and entered production in late spring 1942, strangely after the M4A1 and M4A2 variants. When delivered to the British in October 1942, they called it the General Sherman and the name 'Sherman' stuck. Of simple and uncomplicated design, the M4 was suitable for rapid mass production. Fast, reliable and rugged, it became the most numerous Western Allied battle tank of World War II, with a staggering 70,511 vehicles of all variants being built by the conflict's end in August 1945.

The M4 vehicle was essentially an M3 chassis with the same VVS Suspension, a rear petrol (gasoline) engine, a front wheel drive-sprocket and a welded hull with a three-piece nose. The M4 tank was powered by an up-rated 260kW (350hp) nine-cylinder Wright/Continental R975-C1 air-cooled radial petrol (gasoline) engine, which was a modified aircraft power plant. The Sherman also sported controlled differential steering. The tank had a five-man crew (commander, gunner, loader, driver and co-driver/hull gunner) and weighed 29.9 tonnes (29.4 tons), combat loaded.

The M4 had a top road speed of 39km/h (24mph) and 32km/h (20mph) cross-country, plus a cruising range of 193km (120 miles).

M4 Sherman
Crew: 5
Production: 1942–44
Weight: 29.9 tonnes (29.4 tons)
Dimensions: (L,W,H): 5.82m (19ft 2in) x 2.62m (8ft 7in) x 2.74m (9ft)
Engine: 260kW (350hp) super up-rated Wright/Continental R975-C1 petrol (gasoline)
Road speed: 39km/h (24mph)
Range: 193km (120 miles)
Armament: 75mm (2.95in) M3; 1 x 12.7mm (0.5in) M2HB HMG; 2 x 7.62mm (0.3in) M1919A4 MG
Armour: 12.7–89mm (0.5–3.5in)

M4 Sherman
This mid-production M4A1 Sherman, named 'Major Jim', sports a 12.7mm (0.5in) Browning M2HB heavy machine gun on its rear turret roof during the 1943 Tunisian Campaign.

MEDIUM TANKS

M4A1 76mm(W) Sherman
This USMC Sherman's turret, long barrel and slightly curved frontal glacis plate identify it as an M4A1 76mm(W); it is serving with the 11th Tank Battalion, 10th Armoured Division, during the Winter 1944–45 Battle of the Bulge.

Main armament
The M4 tank's main armament was a short-barrelled, low-velocity 75mm (2.95in) M3 gun in the narrow M34 gun mount. Secondary armament comprised of two 7.62mm (0.3in) Browning MG 1919A4 machine guns (one co-axial, one bow-mounted) and one 12.7mm (0.5in) HB-M2 turret-located heavy anti-aircraft machine gun. The Sherman carried 97 armour piercing and high explosive rounds, plus 4750 7.62mm (0.3in) and 300 12.7mm (0.5in) bullets.

The tank had a maximum armour thickness of 75mm (2.95in) on the turret and 50mm (1.97in) on the hull. It could ford to a depth of 0.91m (3ft) and cross a 2.13m (7ft) ditch.

Later production M4 models possessed a one-piece cast nose and the wider M34A1 gun mount. Very late models had a combination cast/rolled hull front. American firms manufactured 6748 M4s between July 1942 and January 1944.

M4A1
The earliest model to enter production was the M4A1, a slightly-modified M4 to speed up production. The major differences from the M4 were the lack of hull side doors and that it had a cast hull. The prototype was completed in February 1942 and entered production the next month. Very early models carried the 75mm (2.95in) M2 gun counter-weighted due to M3 shortages and two twin fixed hull 7.62mm (0.3in) machine guns.

Later production models dispensed with the machine guns and substituted the 75mm (2.95in) M3 main gun. The nose was also changed from three bolted pieces to one cast piece. M34A1 gun mount and sand shields were added to later production models, and a total of 6281 were produced from February 1942 to December 1943.

M4A2
The second model to enter production was the diesel-powered M4A2. Due

---

M4A1(76)W Sherman
Crew: 5
Production: 1944–45
Weight: 33.3 tonnes (32.8 tons)
Dimensions: (L,W,H): 7.5m (24ft 9in) x 2.67m (8ft 9in) x 2.97m (9ft 9in)
Engine: 298kW (400hp) 9-cylinder Wright/Continental R975-C4
Road speed: 46km/h (29mph)
Range: 161km (100 miles)
Armament: 76mm (3in) M1A1, M1A1C or M1A23; 1 x 12.7mm (0.5in) M2HB HMG; 2 x 7.62mm (0.3in) M1919A4 MG
Armour: 12.7–108mm (0.5–4.25in)

MEDIUM TANKS

M4A2 Sherman

This US Marine Corps Sherman M4A2 variant – vehicle 4689 'Adder' – is equipped with hull-rear deep-wading vents for amphibious landing. The straight angular glacis plate differentiated this variant from the M4A1.

to shortages of Wright/Continental petrol engines, its power plant instead consisted of twin 127kW (170hp) General Motors 6-71 diesel engines, which when twinned were designated the GM 6046. The same production changes emerged as with the M4, except the cast/rolled hull was never applied. Very early models also had spoked road wheels, but these were soon replaced by solid disc wheels with embossed spikes. For secondary armament it carried twin fixed bow-mounted 7.62mm (0.3in) machine guns, similar to the M4A1. Some 8053 were built from April 1942 to December 1943, but few served with US forces. However, some 7413 of them were exported via Lend-Lease, including 5041 to the United Kingdom.

M4A3

The M4A3 was the fifth type to enter production. This variant possessed a welded hull and was powered by a 336kW (450hp) Ford GAA V-8 petrol (gasoline) engine, giving a top road speed of 46km/h (29mph). This was the model most favoured by the US Army, and it became its primary variant. Production changes mirrored those in the M4 but with the rolled/cast nose added. Numerous improvements were integrated into production after combat experience during 1943, which included a commander's vision cupola, loader's hatch, better sloped hull front and 'wet' stowage to reduce the risk of fire. Some 3071 were built from June 1942–March 1945.

M4A4

The M4A4 was the fourth model to enter production in July 1942 but only remained in construction until September 1943. Early vehicles had three-piece bolted nose, an M34 gun mount and vision slots, but these were discontinued and a M34A1 gun mount fitted instead. It was powered by a 276kW (370hp) Chrysler A57 WC Multibank petrol (gasoline) engine.

M4A2 Sherman
Crew: 5
Production: 1942–43
Weight: 31.8 tonnes (31.3 tons)
Dimensions: (L,W,H): 5.92m (19ft 5in) x 2.62m (8ft 7in) x 2.74m (9ft)
Engine: General Motors 6046 (twinned 127kW [170hp] GM 6-71) diesel
Road speed: 46km/h (29mph)
Range: 241km (150 miles)
Armament: 75mm (2.95in) M3; 1 x 12.7mm (0.5in) M2HB HMG; 2 x 7.62mm (0.3in) M1919A4 MG
Armour: 12.7–108mm (0.5–4.25in)

MEDIUM TANKS

M4A3 Sherman

This late-production M4A3 served with the USMC 4th Tank Battalion on Iwo Jima, 1945. Its rear-hull deep-wading vents are still attached and its crew have fitted turret-side track links and hull sandbags as improvised protection upgrades.

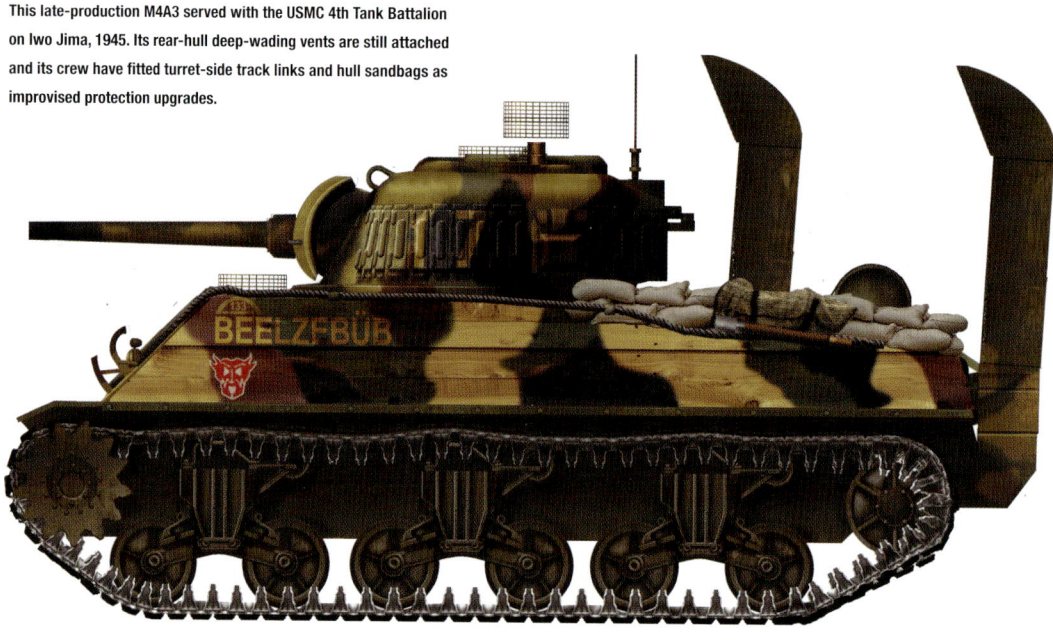

M4A3 Sherman

This 75mm-gunned M4A3, named 'Caballero', served with the 69th Tank Battalion, from the US 6th Armored Division; it is seen here during the 1944–45 Battle for the Ardennes.

MEDIUM TANKS

M4A3 Sherman
Crew: 5
Production: 1942–43
Weight: 30.8 tonnes (30.3 tons)
Dimensions: (L,W,H): 5.92m (19ft 5in) x 2.62m (8ft 7in) x 2.74m (9ft)
Engine: 336kW (450hp) Ford GAA V-8 petrol (gasoline) engine
Road speed: 46km/h (29mph)
Range: 210km (130 miles)
Armament: 75mm (2.95in) M3; 1 x 12.7mm (0.5in) M2HB HMG; 2 x 7.62mm (0.3in) M1919A4 MG
Armour: 12.7–89mm (0.5–3.5in)

This comprised five commercial automobile engines connected to a common drive shaft as another improvised expedient to surmount engine shortages. The M4A4 could obtain a top road speed of 40km/h (25mph). The large parallel engine required an elongated 6.06m (19ft 11in) hull. A total of 7499 were produced.

M4A6
The M4A5 was the US Army designation allotted to but never used for the Canadian Ram tank, also developed directly from the US M3 tank, therefore the diesel-powered M4A6 was the final production model. It had a longer hull and tracks and its power plant comprised a 336kW (450hp) Ordnance RD-1820 nine-cylinder air-cooled radial diesel engine of superior fuel economy. Some 775 were ordered and production began in October 1943 but was discontinued in February 1944 after only 75 vehicles. This discontinuation resulted from the decision to concentrate on petrol engines and the existing M4A3 variant. The M4A6 used late production M4 cast iron hulls, but with larger driver's hatches and a distinctive travel lock for the main armament.

Appliqué armour was welded over the sponson ammunition racks for added protection. Turrets were the late production type with added armour thickness. Pistol ports were initially eliminated but later reinstalled. All carried the modified M34A1 gun mount. Testing proved its superior fuel economy. However, none served

Two Sherman M4A3s of the US 1st Armored Division advance down a dirt road near Anzio, Italy, early 1944; note the mass of kit and equipment the crew has stowed on the rear hull.

overseas but instead remained with testing and training units in the US.

Enhanced firepower
Five major sets of production modifications were introduced during the manufacturing span of the M4 Sherman based upon field combat experience. As the various variants were produced in tandem, these modifications were introduced on most of the production variants in parallel. The major production developments were as follows. Firstly, there were enhanced firepower amendments. The more powerful 76mm (3in) M1 or M1A1 tank gun was incorporated in vehicles produced from February 1944. Such modified Shermans were designated M4A1(76). Secondly, there were enhanced protection upgrades. To reduce the risk of fire, appliqué armour

MEDIUM TANKS

M4A3

M4A3 Sherman Medium Tank 'Doris', D Company, 4th Marine Tank Battalion, Iwo Jima, February 1945.

Front view

This view depicts the design's angular, box-like, hull superstructure; the hull machine gun is visible offset to the right (the viewer's left).

MEDIUM TANKS

Rear view
From the rear, the Sherman's slightly asymmetrically rounded turret can be discerned, as well as the coffin-like angular mass of the hull.

M4A3 Sherman
Crew: 5
Production: 1942–43
Weight: 30.8 tonnes (30.3 tons)
Dimensions: (L,W,H): 5.92m (19ft 5in) x 2.62m (8ft 7in) x 2.74m (9ft)
Engine: 336kW (450hp) Ford GAA V-8 petrol (gasoline) engine
Road speed: 46km/h (29mph)
Range: 210km (130 miles)
Armament: 75mm (2.95in) M3; 1 x 12.7mm (0.5in) M2HB HMG; 2 x 7.62mm (0.3in) M1919A4 MG
Armour: 12.7–89mm (0.5–3.5in)

Overhead view
This view highlights the positioning of the various crew hatches, the central mounting of the external turret-top machine gun, and the location of the various fixtures on the hull rear designed to carry the tools all tank crew required.

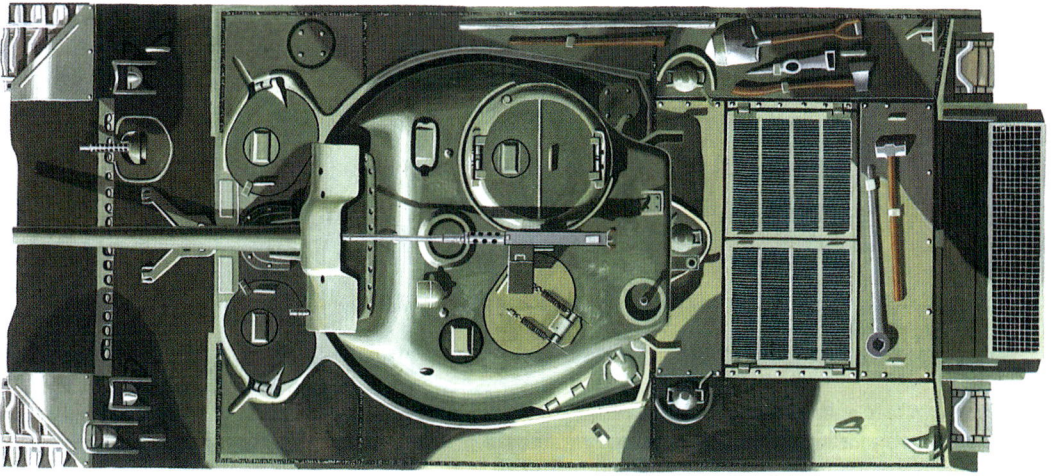

plates and 'wet stowage' (glycerine-protected) ammunition racks were introduced.

Thirdly, there were suspension improvements. To counter the vehicle's increased weight and diminished cross-country performance, wider T80 tracks and the Horizontal Volute Spring Suspension (HVSS) were introduced on some vehicles from mid-1944. Fourthly, American firms developed close support capability variants.

During 1943 a close support Sherman design was developed that mounted a 105mm (4.1in) Howitzer on the M4A2 chassis. These vehicles were given the suffix (105). And finally, there were a range of production simplification modifications introduced. These included a new 47-degree sloped hull front, larger hatches and a new vision cupola.

These numerous modifications led to the emergence of a baffling multitude of new variants. Some 800 M4(105) with a 105mm (4.1in) Howitzer were produced in 1944–45. An additional 841 M4(105) HVSS tanks were built during 1944–45 that featured the howitzer, enhanced protection, HVS Suspension and the production simplifications listed above. Some 3396 M4A1 (76mm/3in) tanks were built during 1944–45 with the more powerful gun, enhanced protection and production simplifications. A further 1615 M4A2 (76mm/3in) were constructed during 1944–45 and another 1925 M4A3 (76mm/3in) in 1944. American firms constructed a total of 1455 M4A3 (76mm/3in) HVSS vehicles with upgraded suspension during 1944. Some 500 M4A3(105) tanks were also built during 1944 as well as an additional 2539 examples of the M4A3(105) HVSS with Howitzer and enhanced suspension.

Special-purpose variants
American firms also developed dozens of special-purpose M4 variants. The M32 tank recovery vehicle had a fixed turret and a powerful winch and jib crane fitted. The M32B1-B4 variants were modified M4A1-A4 tanks. During

---

M4A3 (105) HVSS Sherman
Crew: 5
Production: 1944–45
Weight: 31.5 tonnes (31 tons)
Dimensions: (L,W,H): 5.89m (19ft 4in) x 2.62m (8ft 7in) x 2.74m (9ft)
Engine: 298kW (400hp) up-rated Wright/Continental R975-C4 petrol (gasoline)
Road speed: 39km/h (24mph)
Range: 161km (100 miles)
Armament: 105mm (4.1in) M4; 1 x 12.7mm (0.5in) M2HB HMG; 1 x 7.62mm (0.3in) M1919A4 MG
Armour: 12.7–108mm (0.5–4.25in)

M4A3 (105) HVSS Sherman
Note the 105mm (4.1in) M4 cannon's short stocky barrel sported by this Sherman close fire support tank. Belonging to the US 6th Armored Division, it is seen here in the Ardennes, late 1944, in winter camouflage of whitewashed patches over its base olive drab.

MEDIUM TANKS

M4A4 Sherman
The US supplied this M4A4 to the Chinese Nationalist government to help resist the brutal Japanese invasion; note the yellow patches on the turret roof and mantlet top, probably an identification marker for 'friendly' aircraft.

1944, with the recovery gear removed, the B32B1 became the M34 prime mover. A few M4s were fitted with dozer blades in 1943 as the M4 Dozer tank. It proved so successful that a purpose-built M1 Dozer blade was manufactured for M4 and M4A1 second series dozers during 1944. The blade assembly could be added or removed from standard M4 and M4A1 battle tanks in the field. The M4 Mobile Assault Bridge designed in 1943 was a double track bridge dropped in place by a few specially modified old M4 tanks. The most famous 'add-on' was the Cullin Hedgerow Cutting Device field improvised in Normandy from German beach defences to cut through the Norman Bocage terrain. Hundreds of Shermans were modified with this device.

Aunt Jemima

Multiple mine-clearing variants were also developed. The T1E1-E6 were disc mine exploding devices that could be hitched to the front of Sherman tanks. The most extensively used was the T1E3 'Aunt Jemima' vehicle that was employed in Normandy. The T2 Flail Tank was the US designation for the British Crab Mk. I Sherman tank with mine flail; the similar T3 was the modified British Crab Mk. II. Small numbers found their way into US service. American forces also improvised a number of flame-throwing tanks based on the M4 chassis, which are discussed in the following 'Sherman M3A3R5 flame-throwing tank' entry. As we shall also see later, four British-produced M4 Crocodile flame-thrower tanks modified with fuel lines that ran over the hull rather than under the vehicle's belly also entered US service.

Several rocket launcher variants of the M4 Sherman also entered operational service. The T34 'Calliope' had 60 separate 117mm (4.6in) rocket launcher tubes in four rows on a frame launcher fitted above the

M4A4 Sherman
Crew: 5
Production: 1942–43
Weight: 31.6 tonnes (31.1 tons)
Dimensions: (L,W,H): 6.06m (19ft 11in) x 2.62m (8ft 7in) x 2.74m (9ft)
Engine: 276kW (370hp) Chrysler A57 WC Multibank petrol (gasoline)
Road speed: 40km/h (25mph)
Range: 161km (100 miles)
Armament: 75mm (2.95in) M3; 1 x 12.7mm (0.5in) M2HB HMG; 1 x 7.62mm (0.3in) M1919A4 MG
Armour: 12.7–76mm (0.5–3in)

turret. It entered service in Normandy in August 1944. The T34E2 variant held 60 183mm (7.2in) rockets in a similar frame launcher, and it entered service in 1945. The T40 'whiz-bang' had 20 183mm (7.2in) rockets in a box frame and was fired hydraulically from inside the tank. It saw limited action during 1944–45. Many developmental and experimental variants were also produced that were designated 'E' for experimental. Very

MEDIUM TANKS

M4A3(76)HVSS Shermans pass parked M4A1(76) VVSS tanks during the 36th Tank Battalion's victory parade in Peine, near Hannover, Germany, May 1945. The longer barrel, muzzle brake, and 'pointed' turret rear identify 76mm-equipped Shermans from the standard 75mm variants.

few ever saw service, although small numbers of M4E9s with modified suspension did see service during 1944–45.

Combat debut

The Sherman tank saw its combat debut in US service during November 1943 in the Pacific as part of the USMC's assault on Tarawa, where it played a major role in securing the island. During 1943, it became the primary battle tank used in the Pacific, serving in independent armoured battalions that reinforced infantry divisions. These units were authorized 64 Shermans (including six 105mm/4.1in Howitzer variants). In the European theatre, the Sherman joined combat in early December 1943 in Tunisia during the Allied Operation Torch invasion of French Northwest Africa. A platoon of M4A1s from the 2nd Armoured Division was committed at Tebourba on 6 December and wiped out the same day. During February 1943, the Sherman-equipped 2nd and 3rd Battalions of the 13th Armoured Regiment of the 1st US Armoured Division clashed disastrously with German tanks during the ill-fated attempt to stop Rommel's thrust through the Faid Pass and to retake Sidi Bou Zid, where both battalions were destroyed 14–15 February.

After the Axis collapse in Tunisia in early May 1943, the M4A1 Sherman tank became standardized in the 1st Armoured Division and was used by the formation throughout the Italian campaign; however, it was increasingly replaced from summer 1944 by late model M4A3 tanks. During 1943, the Sherman became the primary combat tank of all US armoured divisions, some 16 of which were raised during the war. In September 1943 14 of these formations were reorganized as 'light' armoured divisions with 216 Sherman variants (168 M4 tanks, 18 M4105 self-propelled Howitzers and 30 M32 turretless recovery tanks). These all served in the European theatre.

The M4 Sherman acquitted itself reasonably well against the German Panzer IV medium tank. The Sherman's fast power traverse usually meant it got off the first shot and its gyrostabilizer ensured good fire accuracy with a well-trained crew. However, it was outclassed by German Panther and

Tiger tanks, whose frontal glacis plates were impervious to Sherman rounds even at point-blank range. Superior numbers and its nimble cross-country performance often allowed Shermans to outflank and engage heavier Panzers from the flank and rear where they were more vulnerable. Its inadequate firepower and the propensity of the early petrol (gasoline) powered Shermans in particular to catch fire when penetrated by enemy fire were the tank's two most significant weaknesses. The flammability problem was lessened in the diesel vehicles and with the subsequent introduction of wet stowage for the ammunition. In the Pacific theatre, where the Sherman engaged enemy defenders at close range, its limited protection proved its greatest weakness. The introduction of the 76mm (3in) gun, appliqué armour, wet stowage, wider tracks and HVS suspension during 1944–45 combined to enhance the Sherman's combat effectiveness vis-à-vis the German tanks they encountered.

British Shermans

Some 20,568 M4s were delivered to the British via Lend-Lease. In British service the tank was named the General Sherman and the Sherman name became synonymous with the tank. The first 300 all-cast hull M4A1s joined the Eighth Army in North Africa in October 1942 in time for the Battle of El Alamein. The M4A4 became the most numerous Sherman in British service with 7167 supplied, followed by the M4A2 (5041 delivered), M4 (2096) and the M4A1 (942); seven M4A3 tanks and specialized variants comprised the remaining 5315 vehicles. The British gave the five main US variants numerical designations, Mk. I–V. Some 593 American-built 105mm (4.1in) Howitzer variants also entered British service with the designator 'A' and a further 1335 76mm (3in)-armed Shermans received the designator 'B'.

The major British independent development was the more powerful 17-pdr (76mm/3in)-armed Sherman Firefly, which received the 'C' designator, and entered production in February 1944. A 'Y' designation was given to variants with wet stowage and HVSS upgrades. The Guards Armoured Division field modified the Sherman V Rocket variant fitted with twin 27kg (60lb) Typhoon aircraft rocket launchers mounted astride the turret. Specialized British Shermans included ARVs (Armoured Recovery Vehicles) and BARVs (Beach Armoured Recovery Vehicle), 52 of which were employed on

M4A6 Sherman

Just 75 elongated diesel-powered M4A6s, the final Sherman variant, were built, and these all remained at US test facilities – hence the rear hull side logo. Note the slightly larger spaces between the bogied road-wheel pairings.

MEDIUM TANKS

Sherman VC Firefly
To augment their Shermans' lethality, the British fitted a modified 17-pounder gun into an adapted turret to create the 'Firefly', one of which served in each four-tank troop. The longer barrel and conspicuous flash on firing, however, made the tank easily identifiable to the enemy.

D-Day. Some 75 Sherman Kangaroos were field converted via turret removal to form improvised APCs. The British 79th Armoured Division developed many specialized Shermans, including Fascine Carriers, OP Command tanks, the turretless Sherman Plymouth Bailey bridge carrier and the Twaby Ark bridging vehicle.

The most famous British modification was the Sherman DD (Duplex Drive) swimming amphibious assault tank employed on D-Day. The Duplex drive was mounted in a waterproofed Sherman fitted with a collapsible canvas float screen. The DD drive was capable of 7.4km/h (4mph) in shallow, calm water. Of high maintenance, many were easily swamped and lost. Most survivors were field converted back to conventional combat tanks. The most effective mine clearers were the Sherman Crab developed in mid-1943. Its flail rotor had 43 flailing chains that beat the ground ahead of the tank detonating mines.

The Sherman Crocodile flame-thrower carried its flame fuel in a towed armoured trailer. Britain developed the Crocodile for the US who ordered 100 Crocodiles in February 1944. However, only four were completed and delivered to the 739th (US) Tank Battalion (Special Mine Exploder Unit). They were first employed during the 29th US Infantry Division's 24 February 1945 assault on Julich.

The Sherman Badger was a 1945 Canadian improvisation mounting a Wasp flamethrower in place of the hull machine gun on a few M4A2 HVSS tanks.

Free French Shermans

Sherman tanks first served in the Fighting French Tank Company in the British Eighth Army at El Alamein and later in Tunisia. During the summer of 1943 the unit expanded into the 2nd French Armoured Division. The Free French received 656 Lend-Lease Shermans during 1943–1944 (382 new M4A2s and 274 M4A4s). With these, the Free French formed three armoured divisions with 165 Shermans each. French field modifications included additional appliqué armour, rear hull stowage boxes and occasionally smoke mortars.

The French 2nd Armoured Division participated in the 1944 Normandy Campaign under Third US Army. It suffered heavy losses liberating Paris during late August 1944 and was

Sherman VC Firefly
Crew: 5
Production: 1942–44
Weight: 37.8 tonnes (37.2 tons)
Dimensions: (L,W,H): 7.85m (25ft 9in) x 2.67m (8ft 9in) x 2.74m (8ft 11in)
Engine: 350kW (470hp) up-rated Chrysler A57 multib-ank petrol (gasoline)
Road speed: 40km/h (25mph)
Range: 201km (125 miles)
Armament: 17pdr (76.2mm/3in) QF gun; 1 x 7.62mm (0.3in) Browning M1919A4 machine gun
Armour: 15–100mm (0.59–3.9in))

replenished from US stocks, including a few M4 105mm (4.1in) tanks, 25 M4A3 (76mm/3in) tanks, 20 new M4A3 (76mm/3in) W vehicles with wider T80 tracks and six old M4 artillery observation vehicles.

During 1944–45, the French 2nd Armoured received various surplus US Shermans as replacements, including M4(105) Howitzers, early-production M4A1s, M4(105) HVSS, stripped-down M4A1 DD tanks and T3 mine flail tanks. The French 1st Armoured Division landed in Southern France in mid-August 1944, largely equipped with M4A4s. The division also received US replacement Shermans, including M4A1, M4A2 and M4A1 (76mm/3in)

MEDIUM TANKS

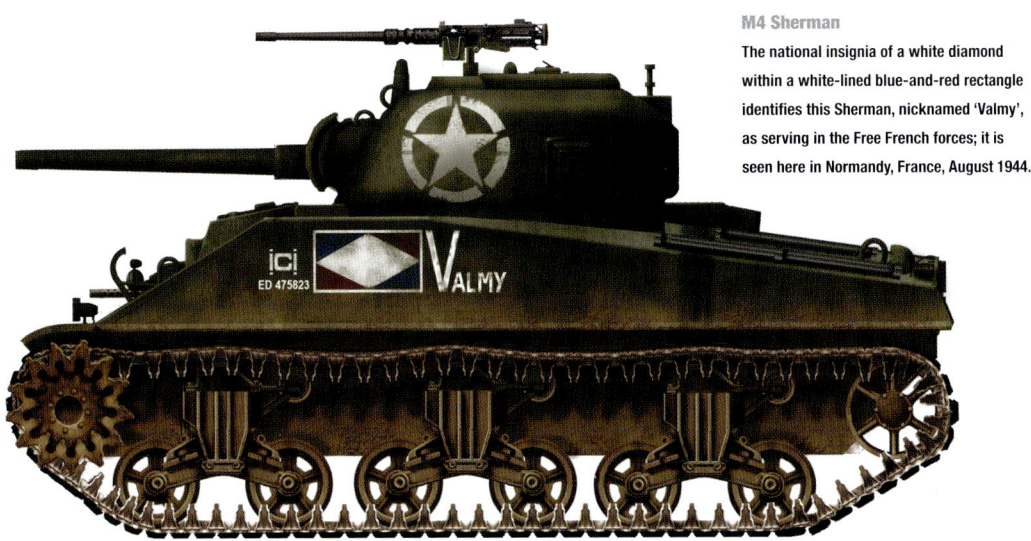

M4 Sherman
The national insignia of a white diamond within a white-lined blue-and-red rectangle identifies this Sherman, nicknamed 'Valmy', as serving in the Free French forces; it is seen here in Normandy, France, August 1944.

M4 Sherman Duplex Drive (DD)
This side view of a Sherman Duplex Drive (DD) amphibious tank nicely depicts, positioned around the hull, its canvas flotation screen in its lowered condition; when fully erected the top of the screen would be just a few inches higher than the turret's roof.

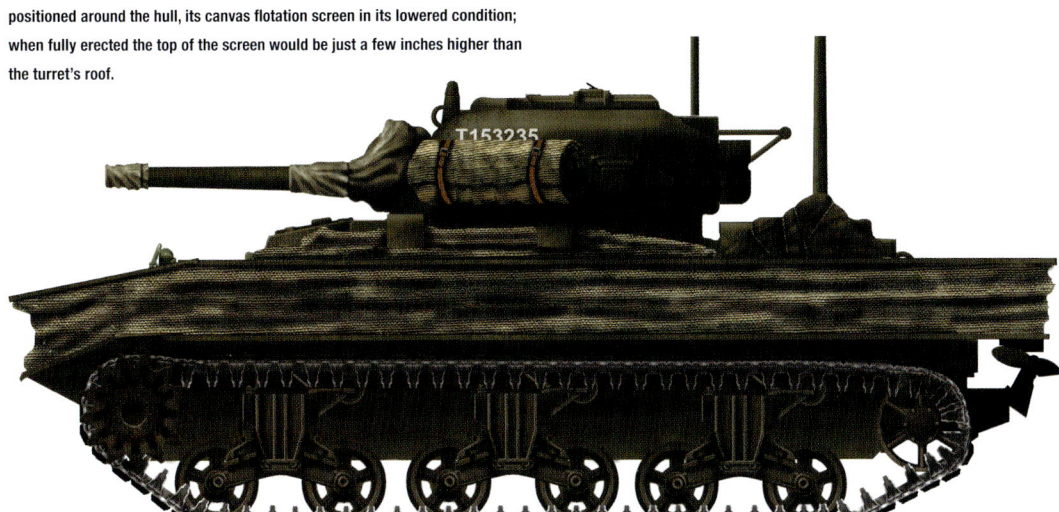

tanks, as well as a few M4(105) HVSS. The 5th Armoured Division joined combat during November 1944, again predominantly equipped with M4A4s. Replacements received included the M4A1 and M4A1 (76mm/3in), as well as a few remanufactured M4A1s with E9 suspension. Other replacements included stripped British Sherman V DD tanks, M4A2s and M4A4s (all identifiable as originally British by their glacis spare track holders and external fire extinguishers) that had been transferred to the US and thereafter passed on to the French. Additionally, Lend-Lease furnished the Soviet Union with 1991 M4A2s and 2 M4A4s, while Brazil received 54 M4s.

MEDIUM TANKS

Sherman Mk. V / M4A4 Crab Mk. I

The Sherman Crab was the result of extensive research by British engineers and designers who operated with limited resources. The Crab was developed in time to accompany British assault troops ashore during the D-Day landings.

Chain

A total of 43 lengths of heavy chain were attached to the Crab's rotating bobbin. Replacement chains were carried aboard the tank in a rack attached to the hull. Cutter disks were added to the flail to destroy barbed-wire entanglements.

Sherman Mk. V / M4A4 Crab Mk. I
Crew: 5
Production: 1942–45
Weight: 31.8 tonnes (31.3 tons)
Dimensions: (L,W,H): 6.35m (20ft 10in) x 2.81m (9ft 3in) x 3.96m (13ft)
Engine: 276kW (370 hp) Chrysler A57 WC Multibank petrol (gasoline)
Road speed: 40km/h (25mph)
Range: 161km (100 miles)
Armament: 75mm (2.95in) gun
Armour: 12.7–76mm (0.5–3in)

M4A3R5 Flame-thrower Tank

After incurring high casualties engaging Japanese bunkers on Saipan during June–July 1944, the newly-created Flame Thrower Group (FTG) began modifying USMC Shermans to produce specialized flame-throwing tanks that troops soon dubbed 'Zippo' after the famous lighter.

On Saipan, American infantrymen using hand-held flame-throwers had shown the stark lethal and psychological effect these weapons had on even fanatical enemy troops. During September 1944, therefore, FTG personnel mounted the CB-H1 flame-thrower in eight diesel-powered Shermans to create the M4A3R5 (CWS-POA-H1) vehicle.

The CB-H1 flame device sported a 75mm (2.95in) projection barrel, which was mounted co-axially alongside the Sherman's main armament. The CB-H1 spewed flame up to a maximum range of 120m (393ft).

The Seabees next fitted the improved CB-H2 thrower, with a 150m (492ft) maximum range, in four M3A3 Shermans to produce the M4A3R5 (CWS-POA-H2) flame-throwing tank. These 12 tanks fought in the February–March 1945 Iwo Jima campaign where they proved effective in dealing with enemy troops holed up in caves and bunkers.

In November 1944 the FTG also converted 54 of the 713rd Tank Battalion's Shermans into M3A3R5 tanks sporting turret-mounted Canadian Ronson F.U.L. Mk. IV flame-throwers. By the war's end in September 1945, the FTG's Seabees had constructed, across these three different designs, 354 M4A3R5 Sherman flame-throwing tanks. In addition, US firms fitted M3-4-3 bow-mounted flame-throwers to 1784 Shermans across all seven main types.

M4A3R5 (CWS-POA-H1) Flame-thrower
Crew: 5
Production: 1944
Weight: 30.3 tonnes (29.8 tons)
Dimensions: (L,W,H): 5.9m (19ft 4in) x 2.62m (8ft 7in) x 2.74m (9ft)
Engine: 336kW (450hp) Ford GAA petrol (gasoline)
Road speed: 46km/h (29mph)
Range: 210km (130 miles)
Armament: 75mm (2.95in) M2; 1 x CB-H1 flame-thrower; 1 x 7.62mm (0.3in) M1919A4 MG
Armour: 12.7–89mm (0.5–3.5in)

M4A3R5

In this improvised flame-throwing tank, the 75mm (2.95in) launcher tube of the flame device was mounted in the gun mantlet co-axially to the right of the tank's 75mm cannon, but just offset slightly higher.

MEDIUM TANKS

USMC M4A2 Flail Tank

During late 1944, mechanics of the USMC's 4th Marine Division, then based in the Hawaiian Islands, constructed a single improvised Sherman Flail Mine-clearing Tank. They did so by taking a diesel-powered M4A2 that had been equipped with a front-located M1 bulldozer device.

The dozer blade connected to two chevroned-shaped rods that were each mounted on the vehicle's second bogie structure. Between the front end of these rods the mechanics fitted a salvaged truck axle. At each end of the axle the mechanics welded a drum-shaped device that had been fitted with 15 individual metal arrays. Each array consisted of a twisted U-shaped metal cable, to which was attached a chain. An improvised drive shaft ran from the axle device through the glacis plate to a salvaged jeep transmission, which connected to the tank's own drive shaft. These modifications created an improvised 'flail' mine-clearing device.

Rotation speed

The bow-gunner operated the 'flail' device, raising and lowering it and controlling its rotation speed. As the tank advanced, he lowered the device and set it rotating until the chains smashed into the ground, detonating any buried enemy mines.

This improvised Flail Sherman was allocated to the USMC's 4th Tank Battalion's 2nd Platoon. It subsequently participated in the February 1945 invasion of Iwo Jima. The tank successfully advanced inland, sweeping for mines, until it was heavily damaged by enemy fire and abandoned.

**USMC M4A2
Flail Tank**
Crew: 5
Production: 1944
Weight: 32.8 tonnes (32.2 tons)
Dimensions: (L,W,H): 5.92m (19ft 5in) x 2.62m (8ft 7in) x 2.74m (9ft)
Engine: General Motors 6046 twinned 127kW (170hp) GM 6-71) diesel
Road speed: 39km/h (24mph)
Range: 241km (150 miles)
Armament: 75mm (2.95in) M3; 1 x 12.7mm (0.5in) M2HB HMG; 2 x 7.62mm (0.3in) M1919A4 MG
Armour: 12.7–108mm (0.5–4.25in)

USMC M4A2 Flail Tank
This image of a unique mine-clearing vehicle nicely shows the improvised 'flail' device concocted by welding U-shaped metal arrays to a drum and then attaching chains to them.

MEDIUM TANKS

M4A3E2 'Jumbo' Assault Tank

The petrol-engined Sherman M4A3E2 was an assault tank later nick named 'Jumbo'. During the winter of 1943–44, the army identified the need for an up-armoured tank to engage the enemy's Siegfried Line bunkers along Germany's western border.

With limited time, the Americans simply modified the existing M4A3 Shermans and skipped the normal design testing process. In March 1944 the Army contracted Fisher to build 250 M4A3E2 assault tanks. During April–July 1944 Fisher produced 254 vehicles, and after testing they arrived in Allied-liberated Northwest Europe during September–October 1944.

Vehicle protection

Fisher welded solid 38mm (1.46in) thick armour plates to the front hull upper glacis plate and the hull sides, raising the vehicle's protection to 101mm (4in) on the glacis plate and 76mm (3in) on the hull sides. This additional armour raised the vehicle's weight to 38.1 tonnes (37.5 tons), which significantly stressed its VVS suspension. The tank also featured an up-armoured T26 turret with additional armour added to the M26 mounting to create an enlarged gun mantlet, which housed the 75mm (2.95in) M3 gun. By adding track width extending grousers, the tank's extra weight only raised its ground pressure rating by merely 0.034 bar (0.5psi) to 0.98 bar (14.2psi). With a standard Ford 336kW (450hp) petrol (gasoline) engine, but working through a modified final drive ratio, the M4A3E2 could obtain a top speed of 35km/h (22mph).

M4A3E2 Jumbo Assault Tank
Crew: 5
Production: 1944
Weight: 38.1 tonnes (37.5 tons)
Dimensions: (L,W,H): 7.54m (24ft 8in) x 2.9m (9ft 6in) x 2.95m (9ft 8in)
Engine: 336kW (450hp) Ford GAA petrol (gasoline)
Road speed: 35km/h (22mph)
Range: 161km (100 miles)
Armament: 75mm (2.95in) M3; 1 x 12.7mm (0.5in) M2HB HMG; 2 x 7.62mm (0.3in) M1919A4 MG
Armour: 12.7–152mm (0.4–6in)

M4A3E2 'Jumbo' Assault Tank

This M4A3E2 was the first vehicle to break through the German siege of Bastogne on 26 December 1944, a feat enshrined by the crew-painted slogan. Note the extra armour on the hull frontal glacis plate and on the gun mantlet.

MEDIUM TANKS

M7

During 1941, the army intended to replace the M3/M5 light tank, but instead developed the T7 – a light tank with the firepower and protection of a medium tank. During 1941–42, six prototype T7 variants were produced: the T7 and T7E1-T7E5.

The welded T7 chassis was based on a VVS suspension like that of the M3/M5 Stuart. Its turret was similar to the M3 Lee and sported the latter's 37mm (1.46in) M6 cannon. The riveted T7E1 featured a redesigned larger turret. The T7E2 was powered by a 254kW (340hp) Wright/Continental R975-EC2 engine and was armed with a British-made 57mm (2.2in) QF 6-pounder Mk. III gun.

The T7E3 and T7E4 sported the same welded turret and hull, yet the latter was powered by twin Cadillac engines while the former had twin Hercules DRXBS diesel engines.

T7 prototype

This T7 was the first ever prototype vehicle of what would become the M7 medium tank. It sports two distinctive VVSS bogie arrangements per side and a turret similar to that of the M3 Lee/Grant.

IHC prototypes

The five-crew T7E5, like the T7E2, mounted the M3 75mm (2.95in) gun of the M3 Lee and M4 Sherman, as well as two 7.62mm (0.3in) Browning machine guns.

This variant became the baseline for the M7 medium tank. The first International Harvester Corporation (IHC) M7 prototypes weighed 24.7 tonnes (24.3 tons), featured 13–64mm (0.5–2.5in) armour and could reach a top speed of 48km/h (30mph). Trials showed that the M7 was too heavy to have a light tank's manoeuvrability but too light to have a medium tank's protection, and the project was thus abandoned.

M7
Crew: 5
Production: 1942
Weight: 24.7 tonnes (24.3 tons)
Dimensions: (L,W,H): 5.23m (17ft 2in) x 2.84m (9ft 4in) x 2.41m (7ft 11in)
Engine: 254kW (340hp) Wright/Continental R975-EC2 petrol (gasoline)
Road speed: 48km/h (30mph)
Range: 161km (100 miles)
Armament: 75mm (2.95in) M3; 2 x 7.62mm (0.3in) M1919A4 MG
Armour: 12.7–64mm (0.5–2.5in)

T23

During 1942, as Sherman production unfolded, the Army began developing a successor medium tank. The first T23 prototype was a lower-silhouetted, more compact vehicle than the Sherman that mounted the longer-barrelled M1A1 76mm (3in) cannon.

The tank featured a Horizontal Volute Spring Suspension (HVSS) and a novel electrical transmission that was married to the 353kW (474hp) Ford GAN V-8 engine. These arrangements enabled the tank to achieve an impressive top speed of 56km/h (35mph). Tests revealed, however, that key modifications to the gun arrangement were needed. Therefore, the second T23 prototype was finished that mounted a slightly shorter version of the M1A1 in a T29 gun carriage mounting within a redesigned turret.

Test model

Tests showed these alterations to be effective. Thus, between November 1943 and December 1945 the Detroit Arsenal manufactured 250 T23s. Weighing 34.2 tonnes (33.7 tons), the five-crew early production vehicle mounted the 76mm (3in) M1 gun, together with one 12.7mm (0.5in) Browning M1-HB heavy and two 7.62mm (0.3in) M1919A4 machine guns. Later vehicles mounted the 76mm (3in) M1A1 gun.

Poor weight distribution

Protected with 12.7–89mm (0.5–3.5in) armoured plates, it was powered by a 336kW (450hp) Ford GAA engine. Military exercises with these vehicles, however, showed that the design had poor weight distribution, and thus it never entered full-scale production.

T23 prototype

In terms of its external appearance, the T23 was atypical of the standard American wartime design practice. The tank featured a large turret with rounded edges positioned well forward on a shallow hull superstructure.

T23 Prototype
Crew: 5
Production: 1943–45
Weight: 34.2 tonnes (33.7 tons)
Dimensions: (L,W,H): 6.02m (19ft 9in) x 3.12m (10ft 3in) x 2.51m (8ft 3in)
Engine: 353 (473hp) Ford GAN petrol (gasoline)
Road speed: 56km/h (35mph)
Range: 161km (100 miles)
Armament: 76mm (3in) M1 or M1A1; 1 x 12.7mm (0.5in) M2HB HMG; 2 x 7.62mm (0.3in) M1919A4 MG
Armour: 12.7–89mm (0.5–3.5in)

HEAVY TANKS

HEAVY TANKS

During 1939–40, the US Army possessed no operational heavy tanks and had limited experience developing such vehicles. Nonetheless, during 1942–43 it developed the M6 heavy tank, but production was soon halted to focus resources on M4 Sherman mass production. Subsequent development led to the deployment in early 1945 of the effective M26 Pershing, later improved upon in the T26E4 Super Pershing.

The following tanks are featured in this chapter:
- M6 Heavy Tank
- M6A2E1
- M26 Pershing
- T26E4 Super Pershing

This tank, one of the first M26 Pershing (T26E3s) to join combat in the European Theatre, fielded by the 14th Tank Battalion, is about to cross the Rhine on 12 March 45 using a ferry operated by the 86th Engineer Heavy Pontoon Battalion; note the collapsed German Ludendorff bridge in the background.

HEAVY TANKS

M6 Heavy Tank

In the years leading up to the start of World War II, American heavy tank development languished in the doldrums. The rapid German defeat of France in the 1940 Western campaign, however, triggered the US Army to begin initial development in the 50-tonne (49.2-ton) heavy tank class.

American firms produced the first experimental vehicle, designated the T1, in the summer of 1940 and it was subjected to extensive field trials during 1941. The multi-turreted T1 featured a cast hull, a Hydra-matic transmission and a 75mm (2.95in) gun as its main armament, together with an auxiliary 37mm (1.46in) cannon.

Wright/Continental

Subsequently, a second prototype – the T1E1 – was developed during 1942. Powered by the potent 634kW (850hp) Wright/Continental G-200 nine-cylinder petrol (gasoline) engine, the T1E1 incorporated the same VVS Suspension as used on the M3 tank, but with four bogie structures per side, each with two small but double-width road wheels; the extra wide tracks helped reduce the heavy vehicle's ground pressure rating. The tracks were protected by a distinctively-shaped side skirt with a horizontal top edge but a regularly undulating bottom edge that merely concealed the top of the road wheels. The T1E1 sported a frontally-located three-crew turret that mounted a 76mm (3in) M7 cannon and a co-axial 37mm (1.46in) gun. The prototype also sported in its bow a ball-mounting for twinned 12.7mm (0.5in) Browning M2HB heavy machine guns. The vehicle was protected by armour plates that ranged in thickness from 44mm (1.73in) to 83mm (3.6in).

Cast or welded?

The first T1E1 produced featured a cast hull, while the second had a welded hull. During 1942, some 20 slightly-modified cast-hulled T1E1s were manufactured that featured electric

M6
Crew: 5
Production: 1943
Weight: 56.5 tonnes (55.6 tons)
Dimensions (L,W,H): 7.54m (24ft 9in) x 3.12m (10ft 3in) x 3m (9ft 10in)
Engine: 634kW (850hp) Wright/Continental G-200 petrol (gasoline)
Road speed: 35km/h (22mph)
Range: 161km (100 miles)
Armament: 76mm (3in) M7; 37mm (1.46in) M6; 2 x 12.7mm (0.5in) M2HB HMG; 2 x 7.62mm (0.3in) M1919A4 MG
Armour: 25–83mm (1–3.6in)

M6 Heavy Tank

This M6, 'Maud', is seen in a two-tone brown and olive camouflage scheme in Italy, 1944; very few of the 16 vehicles manufactured actually made it to combat theatres.

HEAVY TANKS

M6 Heavy Tank
Seen here at the Fort Benning training facility in 1942, this M6 nicely shows the design's unusual one-piece side-skirting arrangement.

transmissions. At the same, time eight pre-production welded-hull T1E2 (M6) vehicles that additionally featured a torque converter to their transmissions were also produced and subjected to trials. In November 1942 the US Army signed contracts for the manufacture of 115 T1E3 tanks, later re-designated M6. Simultaneously, the Army discussed additional contracts for a further 115 M6s to be sent to THE Allies under the Lend-Lease programme.

Weighing 56.5 tonnes (55.6 tons), the series M6 heavy tank had a crew of five: commander, driver, assistant driver, loader and gunner. The tank featured a long low hull superstructure with slightly inwardly-curving sides and a sloped glacis plate. Despite its powerful power plant, the M6 could only manage a top speed of 35km/h (22mph) by road and just 15km/h (10mph) across country.

Short run production
After only 16 M6 tanks had been completed, the US Army decided to cancel the production run so that the resources involved could be transferred to increase the mass production of the M4 Sherman.

In December 1944 the Army declared the M6 to be surplus to requirements and all but three vehicles were scrapped.

In this left-side view on an M6 heavy tank, vehicle 603 at Fort Knox, Kentucky, the long barrels of the front hull paired M2HB machine guns and the tall radio antenna mounted on the turret's rear roof are prominent.

HEAVY TANKS

M6A2E1

During 1944, the Americans undertook experimental design work to discover if the potent long-barrelled 90mm (3.54in) T7 cannon could be effectively married to the M6 tank design. This work led to the development of a larger and taller round-edged turret with a distinctive pronounced rear-turret bustle.

M6A2E1

The very large turret on the M6A2E1 was needed to house the 105mm cannon, when fitted onto a standard M6 hull made for a vehicle that looked badly proportioned and had a worryingly high silhouette.

When the even more potent 105mm (4.13in) T5E1 cannon was married to a modified T1E1 chassis, the M6A2E1 was born. With its forward located turret, the T5E1 cannon over-hanged the front of the vehicle by some 90 percent of its hull length; the turret's significant height, moreover, left the design with a high silhouette. Firing the T32 APCBC round, the T5E1 could penetrate an impressive 221mm (8.7in) of vertical steel RHA.

Weight increase

The tank's modified hull incorporated improved better-sloped armour protection, while that on its turret was increased to up to 275mm (10.8in) thickness. When combined, these various modifications pushed the vehicle's weight up to 70.8 tonnes (69.9 tons), making it one of the heaviest tank designs developed. Such massive weight severely limited its tactical manoeuvrability and rendered

M6A2E1
Crew: 5
Production: 1944
Weight: 70.8 tonnes (69.9 tons)
Dimensions: (L,W,H): 8.61m (28ft 3in) x 3.51m (11ft 6in) x 2.77m (9ft 1in)
634kW (850hp) Wright/Continental G-200 petrol (gasoline)
Road speed: 34km/h (21mph)
Range: 161km (100 miles)
Armament: 105mm (4.13in) T5E1; 1 x 7.62mm (0.3in) M1919A4 MG
Armour: 25–275mm (1–10.8in)

it incapable of crossing most existing road bridges. The American Supreme Commander in Europe, General Eisenhower, argued that the vehicle was not suited for combat in his theatre, and the project was cancelled. Some of the design elements featured in the T26E2, however, were incorporated into the subsequent M26 Pershing heavy tank design.

M26 Pershing

The US Army's late-war M26 Pershing heavy tank can trace its developmental origins through a series of medium and heavy tank prototypes as far back as 1936. Throughout much of the war, American heavy tank development remained in the doldrums, as American industrial efforts focussed on mass-producing the generally effective multi-role M4 Sherman series of medium tanks.

M26 Pershing
This M26 (T26E3) was one of only a few to reach the front line in Germany during 1945. Note how spare track links are mounted on the turret side, providing a modicum of extra protection.

By 1944, however, it had become very apparent that a heavier vehicle with enhanced firepower and survivability was required to counter the newest generation of German AFVs, such as the Panther and King Tiger.

An early developmental precursor of the Pershing was the T20 prototype medium tank, developed in 1942. This low silhouetted and compact vehicle featured a Horizontal Volute Spring Suspension (HVSS), as subsequently fitted to late-production Sherman tanks. The T20 was powered by the 353kW (474hp) Ford GAN V-8 engine that was combined with a rear novel Torqmatic transmission and rear-located sprocket driving wheel.

The prototype featured the longer-barrelled 75mm (2.95in) M1A1 cannon and was protected by 76mm (3in) frontal armour. Due to problems with the Torqmatic suspension, the successor designs – the T22 and T23 – reverted to the standard M4 Sherman suspension arrangements.

Up-gunned prototype

The T25 represented a significantly up-gunned prototype. The tank featured a VSS Suspension and incorporated a new, larger cast turret based on the one sported by the T23. The large turret enabled the tank to mount the potent 90mm (3.54in) cannon. The modified T26 design sported a torsion-bar suspension and heavier armour that had a maximum thickness of 102mm (4in) on its frontal glacis place. The combination of larger turret, heavier gun and thicker armour pushed the weight of the T26 to 46.1 tonnes (45.4 tons); this meant that the T26 was re-classified as a heavy tank.

A slightly modified T26E1 followed, which quickly superseded by the T26E3. After a small pre-production

M26 Pershing
Crew: 5
Production: 1944–45
Weight: 46.1 tonnes (45.4 tons)
Dimensions: (L,W,H): 6.33m (20ft 9in) x 3.51m (11ft 6in) x 2.78m (9ft 1in)
Engine: 336kW (450hp) Ford GAF petrol (gasoline)
Road speed: 34km/h (21mph)
Range: 161km (100 miles)
Armament: 90mm (3.54in) M3; 1 x 12.7mm (0.5in) M2HB HMG; 1 x 7.62mm (0.3in) M1919A4 MG
Armour: 12.7–110mm (0.5–4.5in)

HEAVY TANKS

M26 Pershing
This side view of an M26 nicely depicts the unusual height of the turret roof-top mounting for the Browning M2HB heavy machine gun; note also the distinctively shaped shallow side skirts.

series of 30 T26E3s, the design was standardized as the M26 tank.

From November 1944 onwards, Fischer began producing the main M26 Pershing production run, based upon the T26E3 design. By late February 1945, Fischer had completed 244 Pershing tanks when the Detroit Tank Arsenal joined production. By May, both factories were each producing 200 Pershing tanks per month and by October 1945 some 2212 M26s had been manufactured. The design was named after General of the Armies John J. Pershing, who commanded the American Expeditionary Army in Europe during World War I.

Maximum operating range

The M26 Pershing, which straddled the medium/heavy tank categories, weighed 46.1 tonnes (45.4 tons), and had a crew of five. The tank was powered by a 336kW (450hp) Ford GAF eight-cylinder petrol (gasoline) engine. This gave the tank a power-to-weight figure of 8.8kW/tonne (11.9 hp/ton). The vehicle's running gear comprised six pairs of rubberized medium-sized road wheels, which were connected to an individual wheel-arm and married to a torsion-bar suspension; a rear-located drive sprocket; and a front idler of the same design as the road wheels. The vehicle had wide tracks, thus lowering its ground pressure ratio. The tank could obtain a top speed of 34km/h (21mph) on roads but remained very slow when travelling cross-country. With a full tank of fuel, the vehicle had a modest maximum operating range of 161km (100 miles).

On top of the running gear sat a shallow flat-roofed hull superstructure made of cast sections that were welded together. This hull featured shallow side plates that widened into taller mudguard wedges at the front and back that covered the top of the vehicle's tracks. Situated on the top of the hull's front roof sat a large and tall round-edged turret. Located on the turret's roof was the commander's cupola, which featured six thick glass vision prisms and a periscope.

In terms of main armament, the Pershing mounted the 90mm (3.54 n) M3 L/50 cannon, for which it carried 60 rounds. Firing the M77 Armour-Piercing Round, this potent weapon could penetrate 137mm (5.4in) of vertical RHA at 1000m (1094yds). The Pershing thus had firepower capabilities that surpassed the German Tiger I heavy

M26 Pershing
Crew: 5
Production: 1944–45
Weight: 46.1 tonnes (45.4 tons)
Dimensions: (L,W,H): 6.33m (20ft 9in) x 3.51m (11ft 6in) x 2.78m (9ft 1in)
Engine: 336kW (450hp) Ford GAF petrol (gasoline)
Road speed: 34km/h (21mph)
Range: 161km (100 miles)
Armament: 90mm (3.54in) M3; 1 x 12.7mm (0.5in) M2HB HMG; 1 x 7.62mm (0.3in) M1919A4 MG
Armour: 12.7–110mm (0.5–4.5in)

HEAVY TANKS

tank but remained inferior to that of the King Tiger. This design meant that the long barrel of the M3 cannon over-hung the front of the tank by a considerable degree. Inside the turret, the loader also fired the 7.62mm (0.3in) co-axial Browning machine gun.

The M26 was well protected with armour thickness that ranged from 75mm (2.95in) on parts of the hull sides and up to 102mm (4in) on the hull front. In terms of communication devices, the Pershing's turret mounted an SCR 5-28 radio set, positioned just behind the tank commander. There was also an infantry inter-communication telephone fitted on the back panel of the engine compartment, fixed inside an armoured protective box; this innovation significantly augmented tank-infantry tactical cooperation.

Limited impact

Unfortunately, the promising capabilities of the M26 Pershing only exerted a very limited impact on the Allied advance into Germany as it arrived too late in the war. Although 310 Pershing tanks had been shipped to Europe by the war's end on 8 May 1945, only 22 ever saw combat. These vehicles, from the T26E3 pre-production batch, served in the 3rd and 9th US Armoured Divisions. They joined combat in the Roer River sector during late February 1945. Subsequently, four T26E3s advanced toward the intact bridge over the Rhine at Remagen, which was captured on 7 March 1945.

A very rare sight – a column of nine M26 Pershing tanks parked up by the roadside somewhere in Germany; they are probably moving up to join one of the few front-line units to receive this potent weapon.

M26 Pershing

The design also sported six independent road wheels using a torsion bar suspension rather than the more typical American practice of paired bogie-suspended road wheels.

HEAVY TANKS

M26E4 'Super Pershing'

During 1944, as the American Army was developing the M26 Pershing, discussions were taking place regarding one concern about the new design's effectiveness. This concern was that the Pershing's potent 90mm (3.54in) M3 L/50 cannon was still inferior to the long-barrelled 88mm (3.46in) KwK 43 L/71 gun mounted in the German King Tiger tank.

Consequently, during January 1945 the Americans took an existing M26E1 developmental tank and converted it to mount the formidable long-barrelled T15E1 L/73 version of the 90mm (3.54 in) M3 gun. With a forward-positioned turret, this extremely long-barrelled cannon thus over-hung the vehicle by 50 percent of the hull's length, rendering this tank 8.64m (28 ft 4 in) long.

After trials held in early 1945 the Americans dispatched this vehicle – now redesignated the M26E4 Pilot Prototype No.1 – to Europe for evaluation in real combat against the heaviest German armour that might be encountered.

This M26E4 Pilot No.1 vehicle featured two recuperators fixed on top of the gun that helped manage the weapon's powerful recoil. In addition, this vehicle was easily recognized because it featured a conspicuous external tube-like stabiliser spring positioned above the turret's gun mantlet. Firing a long single-piece round, the potent 90mm (3.5 in) T15E1 cannon could penetrate the Tiger I tank's frontal armour at the impressive range of 3000m (3,281yds). While serving with the US 3rd Armoured Division during spring 1945, concerns grew about the M26E4 Pilot No.1's relatively modest protection, especially its

T26E4 Super Pershing
Crew: 5
Production: 1945
Weight: 48.2 tonnes (47.4 tons)
Dimensions (L,W,H): 8.64m (28ft 4in) x 3.51m (11ft 6in) x 2.78m (9ft 1in)
Engine: 336kW (450hp) Ford GAF petrol (gasoline)
Road speed: 30km/h (19mph)
Range: 161km (100 miles)
Armament: 90mm (3.54in) T15E2; 1 x 12.7mm (0.5in) M2HB HMG; 2 x 7.62mm (0.3in) M1919A4 MG
Armour: 12.7–102mm (0.5–4in)

Barrel length

The sheer length of the M26's 90mm (3.45in) M3 cannon is evident here, as is its double-baffle muzzle brake.

Panther armour

The odd rectangular bloc just in front of the semi-circular curved gun mantlet is the frontal plate of a German Panther tank; added appliqué plates have been added to the upper and lower hull nose surfaces to create improvised spaced armour.

turret mantlet. After all, this tank would be the division's key asset in engaging any enemy King Tiger tanks that might be encountered; it might thus face the potent killing power of the 88mm (3.46in) KwK 43 L/71 cannon. The 3rd Division's workshops therefore welded an 80mm (3.2in) frontal plate salvaged from a knocked-out Panther tank straight onto the turret mantlet; holes were drilled into both sides of this plate so that the tank's gun-sight and co-axial machine gun could still be used as normal.

Improvised appliqué armour

The workshop also welded appliqué plates onto the upper hull to create improvised spaced armour. These field modifications added 3.8 tonnes (3.78 tons) to the vehicle's combat weight.

Pilot No.2

The second prototype M26E4 vehicle, designated Pilot No.2, married a slightly-modified T15E2 gun to a modified T26E3 vehicle; this similarly had a distinctive stabiliser spring mounted above the mantlet. After trials, this vehicle became the basis of the production run, which was redesignated T26E4, but popularly termed the Super Pershing.

Upgunned variant

During March 1945 the Army contracted Fisher to switch the planned production of 1,000 M26 Pershing tanks to the up-gunned M26E4 variant. When the war in Europe ended in early May 1945, the manufacturing run was reduced to just 25 vehicles, which were delivered to the field army during 1946.

T26E4 'Super Pershing'
The T26E4 pilot Prototype No.1 'Super Pershing'.

SELF-PROPELLED GUNS

In addition to tanks, during the 1939–45 war the US Army also developed a range of other fully-tracked self-propelled gun vehicles. These included tank destroyers, lightly-armoured and manoeuvrable AFVs that mounted potent anti-tank cannons; self-propelled guns (SPGs), that mounted indirect-fire artillery pieces such as Howitzers and field guns; and self-propelled anti-aircraft gun vehicles (SPAAGs) for mobile air defence purposes.

The following tanks are featured in this chapter:
- M10 Wolverine tank destroyer
- M18 Hellcat tank destroyer
- M36 tank destroyer
- Howitzer Motor Carriage M8
- 105mm Howitzer Motor Carriage M7 Priest
- 105mm Howitzer Motor Carriage M37
- 155mm Gun Motor Carriage M12
- 155mm Gun Motor Carriage M40
- Multiple Gun Motor Carriage M1

An M10 from the US 701st Tank Destroyer Battalion advances along a mud track near Mount Terminale, southwest of Bologna, Italy, March 1945.

SELF-PROPELLED GUNS

M10 Wolverine tank destroyer

During the war, there were many belligerents who developed tank destroyers: AFVs whose primary tactical mission was to locate, engage and destroy enemy tanks. The Americans developed highly-manoeuvrable, lightly-protected and fully-tracked direct-fire AFVs with a fully-rotating turret with a potent anti-tank cannon.

These vehicles served in independent battalions that were either held in reserve or attached to front-line infantry battalions as required. Collectively these AFVs formed the Tank Destroyer Force and their main role was to respond to and break up any significant enemy armoured breakthroughs, ideally by engaging from ambush locations.

SPG designs

During 1941, the Army had relied on its improvised halt-tracked SPG designs for tank-hunting. These included the 75mm (2.95in) Howitzer Motor Carriage T12. By late 1941, however, the Army had recognized that these expedient designs were inadequate. Thus, the Army requested the development of a purpose-designed fully-tracked tank destroyer with high manoeuvrability and a fully-rotating turret. During early 1942 the Ordnance Department designed a prototype utilizing the ubiquitous M4 Sherman chassis. After extensive field trials and further modifications, the design was accepted for mass production as the 3-inch Gun Motor Carriage (GMC) M10; it soon became dubbed the Wolverine.

Between September 1942 and December 1943, Fisher and three Ford factories produced 4993 GMC M10s. This design utilized the chassis of the M4A2 Sherman. Starting in October 1942 and continuing until November 1943, these firms also produced 1403 M10A1 variants, which were based on the M4A3 Sherman chassis. America eventually sent 514 M10s to Britain

3-Inch Motor Gun carriage M10 Wolverine
Crew: 5
Production: 1942–43
Weight: 29.6 tonnes (29.2 tons)
Dimensions: (L,W,H): 5.97m (19ft 7in) x 3.05m (10ft) x 2m (6ft 7in)
Engine: General Motors 6046 (twinned 127kW [170hp] GM 6-71) diesel
Road speed: 50km/h (31mph)
Range: 300km (186 miles)
Armament: 76mm (3in) M7; 1 x 12.7mm (0.5in) M2HB HMG
Armour: 9.5–57mm (0.375–2.25in)

M10 GMC Wolverine
This left-side view of an early-production M10 depicts the six pairs of large bolt heads on its hull superstructure sides, as well as the two similar pairs on the turret sides.

SELF-PROPELLED GUNS

M10 GMC Wolverine

This view of an M10 during late 1944 in a winter camouflage scheme of white-wash randomly applied over olive drab also shows the unusual design of the turret rear.

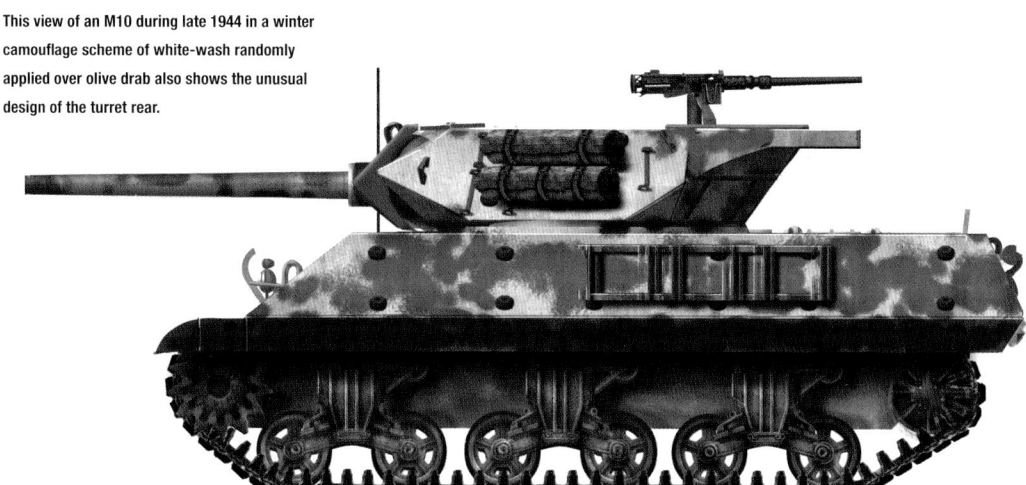

and the USSR under the Lend-Lease Programme.

The squat angular M10 featured the M4 Sherman tank's running-gear: three pairs of medium-sized close-spoked road wheels on a bogie arrangement per side, with a front drive sprocket and rear idler wheel. On top of this sat an angular, well-sloped flat-topped bolted hull superstructure.

Mounted frontally on the hull roof was a large, long open-topped pentagon-shaped angular turret with a distinctive frontal 'beak' and an outwardly-sloped rear face. The turret mounted as the vehicle's main armament was a 76mm (3in) M7 cannon without any barrel muzzle brake. Firing the M79 Armour-Piercing round, at a range of 1000m (1094 yds) the M7 cannon could penetrate 76mm (3in) of RHA sloped at 30 degrees. The M10 carried 54 rounds, the vast majority of which were armour-piercing. These were usually a mix of four types: The M79 Armour Piercing, the M62 Armour Piercing Capped Ballistic Cap,

M93 High Velocity Armour Piercing and M68 Armour Piercing High Explosive rounds. The rear of the turret top edge also had a bracket-mount for a 12.7mm (0.5in) Browning M2HB heavy machine gun, for use in both an anti-aircraft and ground support role; the vehicle carried 300 rounds for this weapon.

Light protection

To save weight and thus boost manoeuvrability, the four-crew M10

An M10, fitted with distinctive deep-wading devices on the hull rear, disembarks onto a beach in Normandy, France, during June 1944.

SELF-PROPELLED GUNS

M10 GMC Wolverine
This M10, 'Siroco', the 621st vehicle produced (by Fisher in January 1943) is seen here in Free French service in Normandy, August 1944.

was relatively lightly protected; its maximum thickness on the turret face was 57mm (2.2in), but elsewhere the plates were just 9.5mm (0.375in) thick. The M10 weighed 29.6 tonnes (29.2 tons), less than the M4 Sherman from which it was derived. The hull and turret's thin bolted steel plates provided greater survivability by being well sloped. The hull sides had six pairs of vertically arranged prominent nuts holding the plates together and the turret sides two such pairs.

The M10 was powered by the 254kW (340hp) General Motors 6046 diesel engine, which paired two GM 6-71 machines. This propulsion enabled the M10 to obtain a maximum road speed of 50km/h (31mph) on roads and an operational range of 300km (186 miles) on a single fuel tank.

Weighing some 0.2 tonnes (0.19 tons) less than the M10, the subsequent M10A1 variant – based on the M4A3 chassis – was hard to distinguish visually from its sister. These M10A1 vehicles were powered by the up-rated 336W (450hp) Ford GAA V-8 petrol (gasoline)

engine, which gave it a 20 percent higher power-to-weight ratio. Both the M10 and the M10A1 served widely in the European Theatre during 1943–45. From late 1944 onwards, however, around 1400 M10s and M10A1s were converted into M36 tank destroyers.

An M10 seems to have moved through a field to take up a firing position on open ground on a slight slope to engage enemy targets near Milan, Italy, in the very last days of World War II.

3-Inch Motor Gun carriage M10 Wolverine
Crew: 5
Production: 1942–43
Weight: 29.6 tonnes (29.2 tons)
Dimensions: (L,W,H): 5.97m (19ft 7in) x 3.05m (10ft) x 2m (6ft 7in)
Engine: General Motors 6046 (twinned 127kW [170hp] GM 6-71) diesel
Road speed: 50km/h (31mph)
Range: 300km (186 miles)
Armament: 76mm (3in) M7; 1 x 12.7mm (0.5in) M2HB HMG
Armour: 9.5–57mm (0.375–2.25in)

M18 Hellcat tank destroyer

Even as the M10 tank destroyer was being developed from 1941 onwards, some senior commanders feared that the vehicle was too heavy and too slow to be fully effective as a tank destroyer. During 1941–42 experimental tank destroyer prototypes were tested to find the tricky optimum balance between high lethality, significant manoeuvrability and adequate survivability.

M18 GMC Hellcat

This M18, 'Amazin Grace', seen in June 1944, sports the design's novel suspension arrangement of five individual large road wheels per side suspended from torsion bars; most US AFVs were fitted instead with bogie-pairings.

In early 1942 a fast tank destroyer, the 37mm Gun Motor Carriage T42, was tested. Sporting a 37mm (1.46in) M3 gun, this prototype was powered by a 220kW (300hp) Wright/Continental R975 petrol (gasoline) engine in the chassis of a Christie-suspension Light Airborne Tank T9. The quest for greater lethality led to the 57mm (2.2in) Gun Motor Carriage T42 prototype, which featured a coil spring suspension instead of the novel Christie one. The sensible decision subsequently to mount the 75mm (2.95in) M3 gun of the early M4 Sherman led to the testing in November 1942 of the 75mm GMC T67.

Further modifications to the main armament and propulsion led to the construction of six 76mm GMC T70 prototypes. This design featured a torsion-bar suspension, a more powerful engine, a redesigned turret, and the 76mm (3in) M1A1 cannon.

Design modifications

After further design modifications the vehicle was accepted for mass production in March 1943 under the designation 76mm (93in) GMC M18; it soon became dubbed the Hellcat by US troops.

Between August 1943 and October 1944 American firms produced a total of 2501 M18s (excluding the six T70 pilot vehicles).

The five-crew M18 was easily distinguishable from its sister M10 and M36 because it did not use the substantial chassis of the M4 Sherman medium tank. Indeed, its smaller lighter chassis brought the Hellcat in at the modest combat weight of 16.8 tonnes (16.5 tons).

76mm Gun Motor Carriage M18 Hellcat
Crew: 5
Production: 1943–44
Weight: 16.8 tonnes (16.5 tons)
Dimensions: (L,W,H): 5.82m (19ft 1in) x 2.87m (9ft 5in) x 2.57m (8ft 5in)
Engine: 260kW (350 hp) Wright/Continental R975-C1 petrol (gasoline)
Road speed: 89km/h (55mph)
Range: 161km (100 miles)
Armament: 76mm (3in) M1A1, M1A1C or M1A2; 1 x 12.7mm (0.5in) M2HB HMG
Armour: 12.7–25mm (0.5–1in)

SELF-PROPELLED GUNS

M18 GMC Hellcat
The muzzle brake fitted to its 76mm M1A1 cannon made a late-production M18 – like this one seen at Anzio during May 1944 – even more visually distinctive from the M10 than earlier vehicles.

Automatic transmission

The M18's running gear comprised five large independent road wheels per side attached to a torsion-bar suspension together with similarly-sized driver and idler wheels and four small return rollers for the upper track. It had a 900T Torqmatic automatic transmission.

The main production vehicle was powered by an up-rated 260kW (350hp) Wright/Continental R975-C1 air-cooled radial petrol (gasoline) engine; late production vehicles received the further up-rated R975-C4 engine that delivered 298kW (400hp). This power plant gave the M18 an impressive top speed of 89km/h (55mph) on tarmacked roads and a creditable off-road speed of up to 43km/h (26mph).

The main production M18 vehicle sported the modified T23 open-topped turret. In its front face the turret featured the 76mm (3in) M1A1 L/52 cannon in an M2 mount, a modification of the standard M1 gun with a greater recoil surface. Later production vehicles

76mm Gun Motor Carriage M18 Hellcat
Crew: 5
Production: 1943–44
Weight: 16.8 tonnes (16.5 tons)
Dimensions: (L,W,H): 5.82m (19ft 1in) x 2.87m (9ft 5in) x 2.57m (8ft 5in)
Engine: 260kW (350 hp) Wright/Continental R975-C1 petrol (gasoline)
Road speed: 89km/h (55mph)
Range: 161km (100 miles)
Armament: 76mm (3in) M1A1, M1A1C or M1A2; 1 x 12.7mm (0.5in) M2HB HMG
Armour: 12.7–25mm (0.5–1in)

The crew of this M18 Hellcat, seen in France, summer 1944, have stowed tarpaulin-wrapped equipment within the metal framework fitted to its turret sides.

SELF-PROPELLED GUNS

instead mounted the M1A1C cannon, which was threaded for a muzzle brake; many of the Hellcats subsequently produced were finished with a muzzle brake fitted on the end of the cannon. Firing an M62 Armour Piercing round, the M1A1 could penetrate 106mm (4.2in) of vertical RHA at a range of 1000m (1094yds).

The combat experiences of 1943–44 exposed the inability of the M1A1 cannon to penetrate the frontal armour of the German Panther tank. This was partly mitigated by the introduction of the M93 High Velocity Armour Piercing round during late 1944. Using its electro-magnetic traverse mechanism, the Hellcat's T23 turret could rotate full circle in just 24 seconds.

On the left rear of the turret was a ring mount that sported a dual-purpose 12.7mm (0.5in) Browning M2HB heavy machine gun. The Hellcat carried 45 main armament rounds, with nine of these available in the turret's right-side container racks.

The M18 design kept its combat weight below 17 tonnes (16.7 tons) largely by merely sporting light armour protection. Most of the hull just had 12.7mm (0.5in) thick plates, although

these were often well-sloped. Even the critical turret front merely featured 25mm (1in) thick plates.

This, in combination with the open-topped turret, rendered the crew highly vulnerable to enemy fire. Engaging the enemy would ideally have to be from well-concealed ambush positions, after which the M18 would 'scoot away' to another pre-prepared firing position.

These issues notwithstanding, the M18 provided sterling service in Europe and the Pacific during 1943–45, being credited with 526 enemy AFV kills.

An early-production M18, with no muzzle brake, tactically co-operates with an infantry squad during the subjugation of Brest, in Brittany, Normandy, September 1944.

M18 GMC

This M18, seen during the winter 1944–45 Battle of the Bulge, was one of the majority such vehicles that sported a 12.7mm (0.5in) Browning M2HB heavy machine gun on its turret roof.

SELF-PROPELLED GUNS

M36 tank destroyer

During 1943 the American Army became aware that the 76mm (3in) M7 cannon mounted on the M10 tank destroyer was not particularly effective against the frontal armour of the German Panther tank unless at close range.

While the 76mm (3in) M1A1 L/52 cannon mounted in the M18 delivered superior lethality, it still struggled to penetrate the frontal armour of the latest enemy tanks at normal combat ranges. This led the Army to develop the potent 90mm (3.54in) M3 L/50 cannon during 1943. This gun delivered impressive lethality – firing the M77 Armour Piercing Round, it could penetrate 137mm (5.4in) of RHA at the typical combat range of 1000m (1094yds). Next, the Americans designed the T71 prototype, which married a new turret that sported the 90mm (3.54in) cannon to a modified chassis of the M10A1 tank destroyer.

After extensive field trials held during fall 1943–44, the design was accepted for mass production under the designation 90mm (3.46in) Gun Motor Carriage (GMC) M36; some authorities allege that troops soon dubbed it the Jackson, after the Confederate US Civil War General. During the period September 1944–June 1945 the firms of ALCO, Fisher, Massey Harris, Montreal Locomotive Works and the Grand Blanc Arsenal produced 1413 M36s, many of which were extant M10A1 vehicles converted to the M36 design through the addition of the new turret.

Armour Piercing High Explosive

Externally, the four-crew M36 resembled the M10A1 but with a taller and longer differently-shaped open-topped turret. This new cast turret had rounded edges and a distinctive large rear bustle that stored eleven 90mm (3.4in) rounds. Indeed, the turret's

---

90mm Gun Motor Carriage M36
Crew: 4
Production: 1944–45
Weight: 29 tonnes (28.5 tons)
Dimensions: (L,W,H): 5.97m (19ft 7in) x 3.05m (10ft) x 3.28m (10ft 9in)
Engine: 336kW (450hp) Ford GAA petrol (gasoline)
Road speed: 42km/h (26mph)
Range: 241km (150 miles)
Armament: 90mm (3.54in) M3; 1 x 12.7mm (0.5in) M2HB HMG
Armour: 12.7–108mm (0.5–4.25in)

GMC M36
This side view of an early-production M36 reveals that its running gear and suspension was almost identical to that of the M10.

SELF-PROPELLED GUNS

GMC M36

In this view of an M36, nicknamed 'Pork-Shop' and finished in monotone olive drab, the new open-roofed turret design and potent 90mm cannon are clearly evident.

length (including bustle) was some 72 per cent of the length of the vehicle's hull roof. On the rear of the bustle there was a bracket-mount that sported a dual-purpose 12.7mm (0.5in) Browning M2HB heavy machine gun, for which the vehicle carried 1000 rounds. The turret featured no co-axial weapon. The M36 carried 47 rounds for its main gun, being a mix of Armour Piercing and Armour Piercing High Explosive rounds. From vehicle 601 onwards, M36s featured an M3 cannon fitted with a distinctive double-baffle muzzle brake.

The chassis and power plant were identical to that of the M10A1. Powered by a 336kW (450hp) Ford GAA V-8 petrol (gasoline) engine, the M36's running-gear sported three pairs of medium-sized close-spoked bogie-suspended VVSS road wheels per side.

The M36 possessed armoured protection similar to that of the M10A1, except that a few critical areas were enhanced. The M36's hull frontal glacis armour was 38–108mm (1.5–4.25in) thick while its turret front had a thickness of 76mm (3in) and its mantlet was 108mm (4.25in). Carrying seven fewer main gun rounds helped compensate for the extra weight of the larger turret and gun and the M36 thus weighed 29 tonnes (28.5 tons), providing it with a power-to-weight ratio of 11.6kW/tonne (11.8hp/ton).

An M36 tank destroyer of the Third US Army trains its gun in the town of Metz, France, during mopping-up operations (22 November 1944).

SELF-PROPELLED GUNS

American firms also produced two M36 sub-variants, designated the M36B1 and M36B2. During mid-1944, the Ordnance Department realised that the rate of M36 production could not meet demand due to shortages of available M10A1 vehicles for conversion. Consequently, between October and December 1944 American firms manufactured 187 M36B1 sub-variants which utilised the chassis of internally-modified Sherman M4A3 tanks. This variant could be differentiated from its sister designs by the 7.62mm (0.3 in) Browning M1919 machine gun mounted on the hull front.

Manufacturing complications

This expedient still did not make up the shortfall in the rate of M36 manufacture, and so in May 1945 American firms produced 52 M36B2 variants. This sub-design was a conversion of the standard M10 (rather than the M10A1) chassis and was thus powered by the 254kW (340hp) General Motors 6046 twin-pair diesel engine instead of the Ford petrol (gasoline) power plant.

Some 672 additional M36B2s were produced after the end of the war in Europe. M36s first entered combat in Europe in October 1944 and eventually replaced the M10 in seven tank destroyer battalions.

The 90mm (3.54in) cannon sported by the M36 proved significantly more lethal than the main weapons mounted in the M10 and M18; M36-equipped battalions regularly recorded long-range destruction of Panther tanks through the employment of enfilade (flank-attack) fire.

90mm Gun Motor Carriage M36
Crew: 4
Production: 1944–45
Weight: 29 tonnes (28.5 tons)
Dimensions: (L,W,H): 5.97m (19ft 7in) x 3.05m (10ft) x 3.28m (10ft 9in)
Engine: 336kW (450hp) Ford GAA petrol (gasoline)
Road speed: 42km/h (26mph)
Range: 241km (150 miles)
Armament: 90mm (3.54in) M3; 1 x 12.7mm (0.5in) M2HB HMG
Armour: 12.7–108mm (0.5–4.25in)

M36 GMC

This late-production M36 in winter camouflage sports an armoured cover over the open-topped turret; the crew have placed sandbags as added protection to both this and the front hull glacis plate.

SELF-PROPELLED GUNS

Howitzer Motor Carriage M8

The US Army also developed a range of fully-tracked self-propelled guns (SPGs) during the 1939–45 conflict. Some of these designs mounted a howitzer, a short-barrelled indirect-fire artillery piece that typically delivered High Explosive rounds in support of ground troops.

One such vehicle was the 75mm (2.95in) Howitzer Motor Carriage (HMC) M8. This was a modified M5 Stuart light tank that mounted a short-barrelled Howitzer in a fully-rotating turret, thus making it the M5's Close Support variant. During mid-1941, the Army developed two prototype 75mm Howitzer Motor Carriage (HMC) T14 SPGs. These fitted a tall box-like superstructure 'fighting compartment' onto the chassis of an M3 Stuart light tank. In the right front of this structure the T14 featured a 75mm (2.95in) Pack Howitzer. Field trials conducted during early 1942, however, showed the design to be unsatisfactory. The SPG was nose-heavy, only had vertical armour plates and its high profile rendered it vulnerable to detection and enemy fire.

Prototype designs

Subsequently, the Army developed a prototype turreted close support version of the M5 light tank, designated the 75mm (2.95in) HMC T41. This design utilized the chassis of the M5 Stuart, onto which was fitted a fighting compartment with the 75mm gun. This experimental design was also found to be excessively nose-heavy. Subsequently, American firms produced a new version of the T41 – designated the T17E1 – that had a manually-operated fully-rotating turret instead of the fixed superstructure. After further testing, the design was accepted into mass production as the 75mm HMC M8. In total, American firms manufactured 1778 HMC M8 SPGs – 373 during late 1942, 1330 during 1943 and 75 vehicles during

---

75mm Howitzer Motor Carriage M8
Crew: 4
Production: 1942–44
Weight: 15.7 tonnes (15.5 tons)
Dimensions: (L,W,H): 4.98m (16ft 4in) x 2.32m (7ft 7in) x 2.75m (9ft)
Engine: twinned 110kW (148hp) Cadillac 44T24 petrol (gasoline)
Road speed: 42km/h (26mph)
Range: 241km (150 miles)
Armament: 75mm (2.95in) M2 or M3; 1 x 12.7mm (0.5in) M2HB HMG
Armour: 9.5–44.5mm (0.375–1.75in)

HMC M8

This view of an HMC M8 shows the very short barrel length of its 75mm L/18.4 howitzer. Based on the M5 Stuart light tank, its compact size is attested to by the relative size of its turret roof-mounted M2HB heavy machine gun.

SELF-PROPELLED GUNS

the final production run in January 1944. Thereafter, the resources devoted to the M8 were reallocated to produce the close support variant of the M4 and M4A3 Sherman tanks – the M4(105) and M4A3(105) – that mounted the 105mm (4.1in) M2A1/M101 L/22 Howitzer.

Based on the standard M5 Stuart chassis, the four-crew lightly-armoured M8 SPG weighed 15.7 tonnes (15.5 tons). On top of the hull roof was mounted a larger rounded-edged cast open-topped turret. On the turret front sat a large bolted curved mantlet set into which via an M7 mounting was the short barrel of the 75mm (2.95in) M2 L/18.4 Howitzer; late-production vehicles mounted the slightly-modified M3 Howitzer. Firing indirectly the M82 High Explosive round with a terminal muzzle velocity of 381m/s (1250ft/s), the Howitzer could obtain a maximum range of 8790m (9610yds). The weapon had a high range of elevation, from -20 to +40 degrees.

Between late 1942 and early 1944, M8s providedl indirect fire support for motorized infantry troops in Europe and the Pacific, but they increasingly became vulnerable to enemy anti-tank fire. During 1944 they were replaced in theatre by the M4(105) and M4A3(105) Sherman close-support tanks.

An M8 engages the enemy from within a small wood; note the amount of equipment the crew have stowed on the turret sides and rear hull.

105mm Howitzer Motor Carriage M7 Priest

During 1941–42, the Americans developed the 105mm (4.1in) HMC T32, a lightly-armoured, fully-tracked, turret-less SPG mounting the 105mm M2A1 L/22 Howitzer in an open-topped fighting compartment on top of the M3 Stuart light tank chassis.

The T32 had tall vertical sides and a sloped front glacis plate, in which sat the howitzer. After trials, the vehicle incorporated a tall ring mount at the superstructure's front side housing a dual-purpose 12.7mm (0.5in) Browning M2HB heavy machine gun; it was from this mounting's Pulpit-like shape that the SPG in British service was dubbed the Priest.

Entering mass production as the 105mm HMC M7 in April 1942, it first entered combat – with the British – during October 1942. Between spring 1942 and early 1944 American firms manufactured 3,749 M7s and 826 M7B1 sub-variants.

Operational distance

Weighing 23 tonnes (22.6 tons), the seven-crew M7 was powered by an up-rated 260kW (350hp) Wright/Continental R975-C1 petrol (gasoline) engine. This enabled it to obtain a top speed of 39km/h (24mph) on roads; on a single fuel tank load it could obtain a maximum operational range of 193km (120 miles). The M7B1 sub-variant utilised the M4A3 Sherman tank chassis – identifiable by the close-spoked rear idler-wheel – and was powered by the 336kW (450hp) Ford GAA V-8 petrol (gasoline) engine. The reliable M7 provided sterling service in Europe and the Pacific through until the war's end.

HMC M7 Priest

On this HMC M7, in Free French service and named 'Franche-Comté', the right-sided drum-like 'pulpit' that housed the mounting for an M2HB heavy machine gun is clearly visible.

105mm Howitzer Motor Carriage M7 Priest
Crew: 7
Production: 1942–44
Weight: 23 tonnes (22.6 tons)
Dimensions: (L,W,H): 6.02m (19ft 9in) x 2.87m (9ft 5in) x 2.95m (9ft 8in)
Engine: 260kW (350hp) up-rated Wright/Continental R975-C1 petrol (gasoline)
Road speed: 39km/h (24mph)
Range: 193km (120 miles)
Armament: 105mm (4.1in) M2A1; 1 x 12.7mm (0.5in) M2HB HMG
Armour: 12.7–108mm (0.5–4.25in)

SELF-PROPELLED GUNS

105mm Howitzer Motor Carriage M37

During 1943–44, the Americans also developed the T76, a new turret-less SPG intended to be more manoeuvrable than the M7. The T76 mounted the 105mm (4.1in) M2A1/M101 L/22 Howitzer as sported by Sherman close-support tanks.

This weapon was housed in a box-like open-topped rear-located fighting compartment fixed upon an extended version of the nimble Light Tank M24 Chaffee chassis, with a torsion-bar suspension. After acceptance for mass production as the 105mm (4.1in) HMC M37 in December 1944, General Motors produced 316 M37s, the first of which reached Europe in May 1945.

Unorthodox gun mounting

Located in the superstructure's centre-right front, the 105mm (4.1 in) M2A1 Howitzer traversed 52 degrees left/right within its cylindrical M5 mounting; the vehicle carried 126 rounds for this weapon. The M37 also featured a distinctive tall cylindrical Pulpit mounting on the superstructure's right front. Atop this sat a dual-purpose 12.7mm (0.5in) Browning M2HB heavy machine gun, for which 550 rounds were carried. Like other American HMCs, the M37 was lightly protected with most plates being no more than 12.7mm (0.5in) thick.

The seven-crew M37 was powered by twin up-rated 164kW (220hp) Cadillac Series 44T4 V-8 petrol (gasoline) engines. At 21 tonnes (20.7 tons) the M37 was 13 per cent heavier than the Chaffee; with a top road speed of 48km/h (30mph) it could not match the Chaffee's manoeuvrability.

HMC M37

A particularly ugly AFV, the HMC M37 mounted a box-like structure on the torsion-bar suspension chassis of the Chafee light tank; its M2HB 'pulpit' was even more prominent than on the M7.

105mm Howitzer Motor Carriage M37
Crew: 7
Production: 1945
Weight: 21 tonnes (20.7 tons)
Dimensions: (L,W,H): 5.49m (18ft) x 3m (9ft 10in) x 2.84m (9ft 4in)
Engine: twinned 110kW (148hp) Cadillac 44T24 petrol (gasoline)
Road: speed 48km/h (30mph)
Range: 161km (100 miles)
Armament: 105mm (4.1in) M2A1; 1 x 12.7mm (0.5in) M2HB HMG
Armour: 9.5–12.7mm (0.375–0.5in)

155mm Gun Motor Carriage M12

During 1941–42, the Ordnance Department developed the 155mm (6.1in) Gun Motor Carriage (GMC) M12, a SPG that mounted a 155mm heavy field gun in an open-topped fighting compartment built upon the chassis of the M3 Lee/Grant medium tank.

During a seven-month production run between September 1942 and March 1943, PSC manufactured 100 155mm GMC M12s. The vehicle had a low open-topped fighting compartment fitted upon the M3 Lee chassis, with the large field gun located at the rear.

The M12 featured either a 155mm (6.1in) M1917, M1917A1 or M1918 M1 L/38.2 field gun, all of which had a maximum range of 19,500m (21,325yds); it featured no other weapons. The SPG carried just 10 rounds for this gun in two compartments located on either side of the fighting compartment's rear. A large rear-fitted earth spade helped absorb the gun's powerful recoil. The Cargo Carrier M30 – which could house 40 rounds of 155mm ammunition – was produced to work alongside the M12. The SPG was lightly protected with most plates being just 12.7mm (0.5in) thick, although key frontal areas had 51mm (2in) thick armour.

The six-crew M12 weighed 27 tonnes (26.6 tons) and was powered by a 260kW (350hp) Wright/Continental R975-C1 petrol (gasoline) engine. It featured a VVS Suspension arrangement with three pairs of bogie wheels. The SPG could achieve a top road speed of 34km/h (21mph).

155mm Gun Motor Carriage M12
Crew: 6
Production: 1942–43
Weight: 27 tonnes (26.6 tons)
Dimensions: (L,W,H): 6.73m (22ft 1in) x 2.68m (8ft 10in) x 2.88m (9ft 5in)
Engine: 260kW (350hp) up-rated Wright/Continental R975-C1 petrol (gasoline)
Road speed: 34m/h (21mph)
Range: 225km (140 miles)
Armament: 155mm (6.1in) M1917, M1917A1 or M1918 M1
Armour: 12.7–51mm (0.5–2in)

GMC M12

The prominent large rear-located earth spade was a distinctive feature of the GMC M12; it helped absorb the potent recoil of the vehicle's 155mm gun.

SELF-PROPELLED GUNS

155mm Gun Motor Carriage M40

The M40 was lightly armoured: most of its welded RHA plates were of 13–25mm (0.5–1in) thickness, although on the lower hull front they were 51–108mm (2–4.25in) thick.

This design solely mounted a long-barrelled 155mm M1A1 or M2 L/45 heavy field gun in an M13 mount within an open-topped rear-located fighting compartment positioned on the widened and lengthened chassis of the M4A3 Sherman. PSC manufactured some 311 M40s before the war's end. One of the first production vehicles served in Europe as a pilot during April 1945. The M2 field gun had a rate of fire of 1.5 rounds per minute and a maximum range of 23,700m (25,919 yds). The gun primarily delivered M112 Armour Piercing and M101 High Explosive rounds, of which the vehicle carried a mix of 20 shells.

With a crew of eight, the M40 weighed 36.3 tonnes (35.7 tons) and was powered by an up-rated 298kW (400hp) Wright/Continental R975-C4 petrol (gasoline) engine. With the M4A3 Sherman's VVSS Bogie-arrangement suspension, the SPG could reach maximum sustained on-road and off-road speeds of 34km/h (21mph) and 23km/h (14mph) respectively.

155mm Gun Motor Carriage M40
Crew: 8
Production: 1945
Weight: 36.3 tonnes (35.7 tons)
Dimensions: (L,W,H): 7.12m (23ft 4in) x 3.15m (10ft 4in) x 3.3m (10ft 10in)
298kW (400hp) Wright/Continental R975-C4 petrol (gasoline)
Road speed: 34km/h (21mph)
Range: 161km (100 miles)
Armament: 155mm (6.1in) M1A1 or M2
Armour: 13–108mm (0.5–4.25in)

GMC M40
This view of a GMC M40 reveals the modified Sherman M4A3 chassis with its Horizontal Volute Spring Suspension (HVSS) rather than the more numerous VVSS version.

SELF-PROPELLED GUNS

Multiple Gun Motor Carriage M19

The final class of fully-tracked SPGs that the US Army developed during the war were self-propelled anti-aircraft guns (SPAAGs). During 1943–44, the Ordnance Department developed the T65E1 prototype SPAAG.

Mass production

The T65E1 prototype sported two 40mm (1.6in) Bofors M2 auto-cannon in a rear-located fully-rotating open-roofed turret mounted on the prototype chassis of the M24 Chaffee light tank. The M2 auto-cannon had a usually-wide elevation range of between -3 and +85 degrees.

The T65E1 prototype was subsequently accepted for mass production as the Multiple Gun Motor Carriage (MGMC) M19 in May 1944. Cadillac manufactured 285 M19s before the war's end and while some vehicles were deployed for Europe they did not participate in combat.

The six-crew M19 carried 352 rounds for its main armament and it could also tow an M28 Ammunition Trailer that carried a further 320 rounds.

Cadillac Series 44T4

The 17.9-tonne (17.6-ton) M19 was lightly armoured with RHA plates of 6.4–12.7mm (0.25–0.5in) thickness, some sloped at 45–59 degrees. The M19 was powered by twinned 148kW (110hp) Cadillac Series 44T4 petrol (gasoline) engines. With running gear of five large road wheels per side on torsion springs, and a twin hydra-matic transmission, this SPAAG could obtain a top road speed of 56km/h (35mph). With a fuel capacity of 416 litres (110gal), the M19 could also obtain a maximum by-road operational range of 161km (100 miles).

Multiple Gun Motor Carriage M19
Crew: 6
Production: 1945
Weight: 17.9 tonnes (17.6 tons)
Dimensions: (L,W,H): 5.81m (19ft 1in) x 2.93m (9ft 7in) x 2.96m (9ft 8in)
Engine: twinned 148kW (110hp) Cadillac 44T4 petrol (gasoline)
Road speed: 56km/h (35mph)
Range: 161km (100 miles)
Armament: two 40mm (1.6in) Bofors M2
Armour: 6.4–12.7mm (0.25–0.5in)

M19
This view of an M19 SPAAG reveals its paired 40mm auto-cannon housed in a fully-rotating rear-located turret; note also the torsion-bar suspension.

AMPHIBIOUS LANDING VEHICLES

During the war, America also developed a series of fully-tracked amphibious landing vehicles (designated LVTs) for service with the USMC, the Navy and the Army; American Allies also adopted them. The early designs merely transported second-echelon troops after the initial assault, but later designs were sufficiently gunned and armoured to serve as genuine amphibious assault vehicles.

The following AFVs are featured in this chapter:

- Landing Vehicle Tracked, Mk. 1 (LVT-1)
- LVT-2 Water Buffalo
- LVT(A)-1
- LVT-3 Bushmaster
- LVT-4 Water Buffalo
- LVT(A)-4
- GMC DUKW

A USMC Squadron of LVT-4 amphibians, with 30 soldiers crowded in their rear compartments, swim through relatively calm seas during the February 1945 American invasion of the Japanese-defended island of Iwo Jima.

AMPHIBIOUS LANDING VEHICLES

Landing Vehicle Tracked, Mk. 1 (LVT-1) Alligator

The design of the first American military amphibious landing vehicle, designated the LVT-1, originated in the civilian world.

Back in 1933, engineer Donald Roebling had developed an amphibious emergency rescue vehicle to be used in Florida's Everglades swamps.

Subsequently, during the summer of 1940 the US Navy and USMC developed and tested the first three LVT-1 military prototypes. These test vehicles were powered by an 89kW (120hp) Lincoln-Zephyr V12 engine. Next, during the second half of 1941, the Food Machinery Corporation (FMC) manufactured the first batch of 200 LVT-1 production vehicles.

These arc-welded steel early production vehicles had tall lozenge-shaped tracks each side fitted around numerous small rollers locked into a rigid suspension that was built around centrally-positioned lozenge-shaped sponsons. The tracks had short aluminium grousers fitted that enabled the amphibious vehicle to move through water at up to 10km/h (6mph). With the ability to have one track in forward gear and one in reverse, the LVT-1 could swiftly turn when moving in water.

The vehicle had a maximum height of 2.33m (7ft 8in) and a width of up to 3m (9ft 10in); typically, it moved through water submerged to the height of the top track, meaning that four-fifths of its mass was typically underwater when in amphibious mode.

On top of the LVT-1's chassis was a low and long superstructure. The vehicle front had a sharp chevron-shaped nose to aid see-worthiness that also featured two lamps. The hull nose ran back and upwards to form the slightly raised cabin area that featured

LVT-1 Alligator
Crew: 3 (+18 troops)
Production: 1941–42
Weight: 14 tonnes (13.8 tons)
Dimensions: (L,W,H): 6.55m (21ft 6in) x 3m (9ft 10in) x 2.48m (8ft 2in)
Engine: 89kW (120 hp) Lincoln-Zephyr
Road speed: 32km/h (20 mph)
Range: 241km (150 miles) on land
Armament: none or 1–2 x 12.7mm (0.5in) M2HB HMG; 1–2 x 7.62mm (0.3in) M1919A4 MG
Armour: none

LVT-1
This LVT-1 is most probably a mid-production example, as it sports some firepower, albeit a solitary 7.62mm (0.3in) Browning M1919A4 machine gun mounted on the rear of the cabin roof.

AMPHIBIOUS LANDING VEHICLES

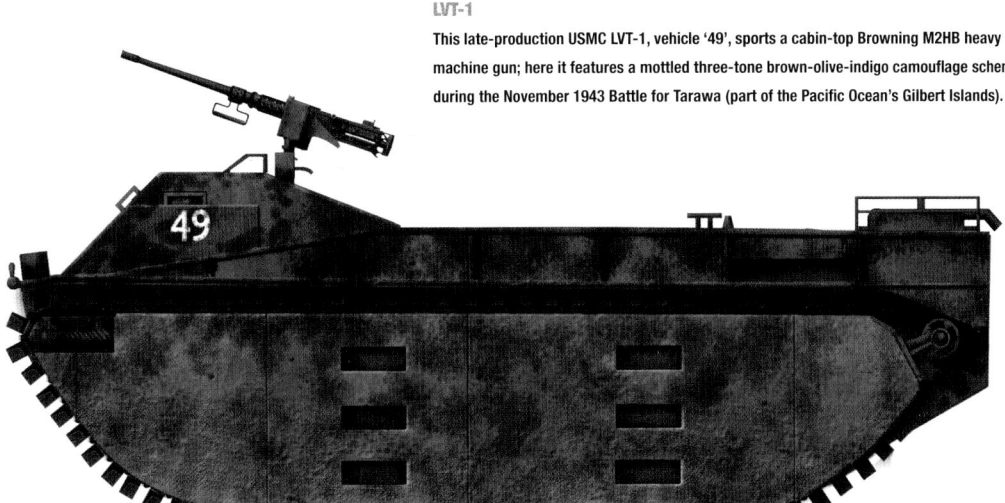

LVT-1
This late-production USMC LVT-1, vehicle '49', sports a cabin-top Browning M2HB heavy machine gun; here it features a mottled three-tone brown-olive-indigo camouflage scheme during the November 1943 Battle for Tarawa (part of the Pacific Ocean's Gilbert Islands).

three evenly-spaced windows. After the cabin, the hull then sloped back to the flat-surfaced rear hull that contained both the engine and the open-topped troop-carrying compartment.

These early LVT-1s weighed 14 tonnes (13.8 tons) and could carry 18 fully-laden troops or 2.1 tonnes (2.07 tons) of cargo. Powered by an up-rated 109kW (146hp) Hercules WXLC-3 engine, the vehicle could obtain a modest land speed of just 15km/h (12mph) across soft terrain. Rather than forming an assault wave, these vehicles were intended to be second-echelon vehicles that would bring cargo or troops ashore once the initial landings had been achieved. They were thus unarmoured and featured no weapons.

First blood

A batch of 128 early LVT-1s first saw combat during the August 1942 Guadalcanal invasion in the Japanese-occupied Solomon Islands where troops soon coined the nick-name Alligator. Later production models featured one or two 12.7mm (0.5in) Browning M2HB heavy machine guns on the cabin roof as well as one or two 7.62mm (0.3in) M1919A4 machine guns on the rear hull; appliqué armour to the cabin front was often added in-theatre.

Combat experience soon revealed that the LVT-1's tracks and suspensions were often unreliable when moving over hard ground. During 1941–42 American firms produced 1225 LVT-1s.

A US Marine Corps LVT-1 churns toward the shore at Cape Torokina, Bougainville, on the first day of landings, 1 November 1943.

AMPHIBIOUS LANDING VEHICLES

LVT-2 Water Buffalo

Combat experience with the LVT-1 proved its concept but also pointed out potential improvements in terms of armour, firepower and seaworthiness. This led to the development of the redesigned LVT-2 Water Buffalo.

LVT-2 Water Buffalo
In comparison with its LVT-1 predecessor, this view of an LVT-2 reveals its shallower, more streamlined hull superstructure and more compact cabin.

There were similarities with the LVT-1's track and suspension, but the lozenge-shaped sponson had been modified and there were differences in the rigid suspension format; the grousers were now bolted-on. Above the tracks, the LVT-2 had only a very shallow, better-streamlined hull superstructure. The now lightly-armoured cabin was smaller and lower than in the LVT-1 and had just two large windows.

At both corners of the cabin roof rear was mounted a 12.7mm (0.5in) M2HB Browning heavy machine gun. There was also a mount at either side of the rear superstructure that could house a 7.62mm (0.3in) M1919A4 machine gun. In some LVT-2s the latter were replaced with additional M2HB weapons. The centrally-located troop-carrying compartment remained open and could house up to 24 fully-laden troops or 2.7 tonnes (2.65 tons) of cargo. Weighing 15.1 tonnes (14.9 tons), the LVT-2 had a more powerful rear-located 186kW (250hp) Wright/Continental W670-9A engine. This increased the vehicle's maximum speed through water slightly to 10km/h (6mph). American firms produced 2963 LVT-2s during 1942–44, and they saw wide service in both the European and Pacific theatres.

US Marines jump over the sides of an LVT-2 as it nears a beach on Kwajalein Atoll, January 1944; later designs incorporated a rear exit ramp that enabled the troops being transported to disembark without exposing themselves unduly to enemy fire.

LVT-2 Water Buffalo
Crew: 3 (+16 troops)
Production: 1942–43
Weight: 15.1 tonnes (14.9 tons)
Dimensions: (L,W,H): 7.95m (26ft 1in) x 3.25m (10ft 8in) x 2.5m (8ft 2in)
Engine: 186kW (250 hp) up-rated Wright/Continental R670-9A petrol (gasoline)
Road speed: 30km/h (19mph)
Range: 241km (150 miles) on land
Armament: 2–3 x 12.7mm (0.5in) M2HB HMG; 1–2 x 7.62mm (0.3in) M1919A4 MG
Armour: None

AMPHIBIOUS LANDING VEHICLES

LVT(A)-1

When the Americans developed an up-armoured LVT variant they added the suffix (A) to indicate this. The first such vehicle, the LVT(A)-1, was an up-armoured fire support variant of the LVT-2. Early amphibious assault experiences indicated that landing vehicles with more potent firepower were required.

This led to the development of the LVT(A)-1 which was similar to a light submersible tank in that it had no troop-carrying capacity. The LVT(A)-1's shallow flat-topped superstructure mounted the same turret as seen on the M3A1 Stuart light tank. This turret contained in an M44 mounting the standard 37mm (1.46in) M6 L/53 cannon and to the right a co-axial 7.62mm (0.3in) M1919A4 machine gun. Behind the turret the rear superstructure dropped slightly to a line just above the top of the tracks. On this flat rear hull were mountings for a further two 7.62mm (0.3in) M1919A4 machine guns each protected by a small shield.

Enhanced survivability upgrades
The plethora of firepower led to an increased crew of six, comprising the

LVT(A)-1
This LVT(A)-1 close fire-support amphibian, a redesigned LVT-2 that carried no troops, mounted an M3 Stuart light tank turret; this vehicle sports an unusual disruptive two-tone camouflage scheme of randomly-shaped olive and khaki patches.

commander, driver and four gunners. The LVT(A)-1 also incorporated enhanced survivability upgrades. The hull decking comprised 6.4mm (0.25in) thick plates while the hull front and turret front had armour that was 51mm (2in) thick.

Its fuel capacity gave the vehicle a maximum operational range of 201km (125 miles). During 1942–44, FMC produced 509 LVT(A)-1s that saw service in the Pacific before being superseded by the more potent LVT(A)-4.

LVT(A)-1
Crew: 6
Production: 1942–44
Weight: 14.9 tonnes (14.7 tons)
Dimensions: (L,W,H): 7.95m (26ft 1in) x 3.25m (10ft 8in) x 3.07m (10ft 1in)
Engine: 95kW (124hp) Hercules WXLC-3 petrol (gasoline)
Road speed: 32km/h (20mph)
Range: 201km (125 miles) on land
Armament: 37mm (1.46in) M6; 3 x 7.62mm (0.3in) M1919A4 MG
Armour: 6.4–51mm (0.25–2in)

AMPHIBIOUS LANDING VEHICLES

LVT-3 Bushmaster

The Borg-Warner Corporation developed the LVT-3 Bushmaster during 1943. Visually, the vehicle looked somewhat like a hybrid mix of the LVT-1 and LVT-2. It had a slightly less shallow hull superstructure than the latter and a cabin more akin to that of the former; the cabin had five armoured windows.

The LVT-3 Bushmaster sported two key innovations. Firstly, it had a bespoke hydraulically-operated exit ramp at its rear that enabled troops to disembark without unduly exposing themselves by jumping over the sides, as in the LVT-1 and -2. Secondly, the LVT-3 had a new powertrain. It mounted paired 110kW (148hp) Cadillac petrol (gasoline) engines, one in each sponson, that were connected through a transmission to a front-located drive-sprocket.

The cabin roof mounted one or two 12.7mm (0.5in) Browning M2HB heavy machine guns. Later production vehicles, with appliqué armour, often had a three-sided gun-shield for these weapons; rails on the troop compartment side also enabled another one or two 7.62mm (0.3in) machine guns to be operated. The Bushmaster had a wider open-topped hold area than its predecessors, which enabled a jeep, as well as troops, to be transported. Its maximum capacity was 30 soldiers or a 4.1-tonne (4-ton) cargo payload. During 1943–45 American firms manufactured 2962 LVT-3s and they first saw combat at Okinawa during April 1945.

LVT-3 Bushmaster
Crew: 3 (+30 troops)
Production: 1943–45
Weight: 17.5 tonnes (17.2 tons)
Dimensions: (L,W,H): 7.45m (24ft 5in) x 3.4m (11ft 2in) x 3.02m (9ft 11in)
Engine: twinned 110kW (148hp) Cadillac 44T24 petrol (gasoline)
Road speed: 27km/h (17mph)
Range: 241km (150 miles) on land
Armament: 1–2 x 12.7mm (0.5in) M2HB HMG; 1–2 x 7.62mm (0.3in) M1919A4 MG
Armour: none or 6.4–12.7mm (0.25–0.5in) appliqué plates

LVT-3 Bushmaster

This late-production LVT-3 is identifiable as such by the three-sided shield for its Browning M2HB heavy machine gun and the horizontal rails along the sides of its rear-located troop-carrying compartment; note also the changes to its sponsons and running gear.

LVT-4 Water Buffalo

The next amphibious landing vehicle to be developed was the LVT-4. Based on the LVT-2, it was designed with a rear-located exit ramp that reduced risk to the troops on disembarkation.

To facilitate this the up-rated 186kW (250hp) Wright/Continental W670-9A engine was relocated behind the crew cabin, thereby facilitating a larger open-topped troop-carrying compartment to the rear. This redesign increased the vehicle's troop-carrying capacity from 16 (in the LVT-2) to 30.

The LVT-4 was visually distinguishable from its sisters in that its less shallow superstructure top was entirely flat; the frontal cabin was not raised above this top line and had three small vision ports rather than large windows.

Weapons

The LVT-4 housed a variety of weapons: either one or two centrally-positioned 12.7mm (0.5in) Browning M2HB heavy machine guns and one or two 7.62mm (0.3in) M1919A4 machine guns; the M2HB weapons were often fitted with a gun shield.

LVT-4 Water Buffalo

This late-production LVT-4, vehicle 'A43', sports a three-sided gun shield for its Browning M2HB heavy machine gun and two-tone camouflage; it participated in the USMC amphibious assault on Yellow Beach at Iwo Jima in February 1945.

FMC, Graham-Paige and the St Louis Car Company between them produced 8348 LVT-4s between December 1943 and September 1945. The LVT-4 saw combat during the 1944 American operations in Saipan, Guam, Tinian and Peleliu.

Some 500 vehicles were transferred to the British Army. The British fitted many of these vehicles with a 20mm (0.78in) Polsten cannon and two 7.62mm (0.3in) Browning machine guns. They participated in the autumn 1944 amphibious assaults in the Scheldt Estuary and the March 1945 attack across the Rhine.

LVT-4 Water Buffalo

Crew: 3 (+30 troops)
Production: 1943–45
Weight: 16.5 tonnes (16.2 tons)
Dimensions: (L,W,H): 7.95m (26ft 1in) x 3.25m (10ft 8in) x 2.5m (8ft 2in)
Engine: 186kW (250hp) up-rated Wright/Continental W670-9A petrol (gasoline)
Road speed: 24km/h (15mph)
Range: 241km (150 miles) on land
Armament: 1–2 x 12.7mm (0.5in) M2HB HMG; 1–2 x 7.62mm (0.3in) M1919A4 MG
Armour: none or 6.4–12.7mm (0.25–0.5in) appliqué plates

AMPHIBIOUS LANDING VEHICLES

LVT(A)-4

Although the LVT(A)-1 had provided much-needed fire support for America's 1944 amphibious assaults in the Pacific, the 37mm (1.46in) M6 cannon was soon deemed to lack sufficient punch. This realisation led to the development by March 1944 of the up-gunned LVT(A)-4.

This 18.3-tonne (18-ton) design featured a low centrally-positioned superstructure. At the rear sat the partially open-topped can-shaped turret of the 75mm (2.95in) Howitzer Gun Carriage M8. This turret featured the short-barrelled 75mm L/18.4 M2 Howitzer in an M7 mounting. The vehicle carried 100 rounds for its main gun, these being a mix of M48 High Explosive and M66 High Explosive Anti-tank shells.

Ball mounting
The rear turret top also sported a ring-mount for a 12.7mm (0.5in) Browning M2HB heavy machine gun. The six-crew vehicle sported 6mm (0.25in) side and rear armour, 6–12.7mm (0.25–0.5 in) thick frontal hull armour and a turret front with a thickness of 38mm (1.46in). Many vehicles also featured a ball mounting in the hull cabin front that housed a 7.62mm (0.3in) M1919A4 machine gun. Sporting in its rear a 186kW (250hp) Wright/Continental W670-9A petrol (gasoline) engine, the LVT(A)-4 married this to a Spicer synchromesh transmission with one reverse and five forward gears.

In the war's last weeks, 269 examples of an improved variant, the LVT(A)-5, entered service. This had a gyro-stabilized M3 gun on the new M12 mounting fitted to an electric powered turret.

LVT(A)-4
From late 1944, the LVT(A)-4 provided the USMC's amphibious assault forces with a much-needed enhanced close fire-support capability – in the form of a short-barrelled 75mm M2 L/18.4 howitzer. These vehicles fought in the USMC's 1945 campaigns, such as Iwo Jima and Okinawa.

LVT(A)-4
Crew: 6
Production: 1944–45
Weight: 18.3 tonnes (18 tons)
Dimensions: (L,W,H): 7.95m (26ft 1in) x 3.25m (10ft 8in) x 3.11 (10ft 2in)
Engine: 186kW (250hp) Wright/Continental W670-9A Petrol (gasoline)
Road speed: 24km/h (15mph)
Range: 241km (150 miles) on land
Armament: 75mm (2.95in) M2; 1 x 12.7mm (0.5in) M2HB HMG; 2–3 x 7.62mm (0.3in) M1919A4 MG
Armour: 6.4–38mm (0.25–1.5in)

GMC DUKW

During 1942, the General Motors Corporation (GMC) developed a wheeled (rather than tracked) amphibious utility vehicle designated by the codename DUKW; in common parlance it became known as 'the Duck'. The vehicle was based on GMC's ACKWX 2-tonne (1.8-ton) cab-over-engine military truck, employing existing design to enable economic mass-production.

GMC welded a water-tight and streamlined all-steel thinly-armoured hull onto this chassis. This rectangular hull had flat sides, a horizontal floor, a curved front nose and an upwardly-curving rear. The DUKW's hull sat upon its six large rubber-edged wheels, with one forward and two rear-located axles.

The front third of the vehicle housed the power plant, a 70kW (94hp) GM-270 six-cylinder engine. The engine powered both the road wheels and – via a long transmission shaft – also the propeller housed at the hull rear. The DUKW could obtain an impressive 80km/h (50mph) on land and 10km/h (6mph) at sea.

The vehicle could travel an incredible 93km (58 miles) across water, making it capable, for example, of crossing the English Channel during the June 1944 Allied invasion of German-occupied Normandy. Behind the engine came the open plywood driver's station. The remaining two-thirds of the boat was filled with its open hold, which could either carry up to 25 troops or a 2.4-tonne (2.3-ton) cargo load.

Between December 1942 and September 1945 GMC manufactured 21,137 DUKWs, some 4000 of which were sent to America's Allies. The DUKW saw extensive combat service in both the European and Pacific theatres.

GMC DUKW
Crew: 5 (plus 25 troops)
Production: 1942–45
Weight: 6.2 tonnes (6.1 tons)
Dimensions: (L,W,H): 9.45m (31ft) x 2.44m (8ft) x 2.69m (8ft 10in)
Engine: 70kW (94hp) General Motors 270
Road speed: 80km/h (50mph)
Range: 644km (400 miles) on land
Armament: none or 1 x 12.7mm (0.5in) M2HB HMG
Armour: none

DUKW
This view of the DUKW, a cheap simple wheeled amphibian intended for economical mass production, nicely shows the distinctive shape of the vehicle's thin hull superstructure.

HALF-TRACKS AND ARMOURED CARS

HALF-TRACKS & ARMOURED CARS

In addition to tanks, destroyers and SPGs, the Americans also designed and deployed armoured half-tracked vehicles and wheeled armoured cars during World War II. The most significant of these included the M2, M3 and M5 half-tracks, the M3 Scout Car, plus the M1 and M18 armoured cars. These vehicles fulfilled the roles of scouting, reconnaissance and the tactical movement of embussed troops across the battle-space.

The following AFVs are featured in this chapter:

- M2 Half-track
- M3 Half-track
- M5 Half-track
- M3 Scout Car
- M1 Armoured Car
- M8 Greyhound Armoured Car

An M8 armoured car weaves its way through parked vehicles in the US forces' rear area during the winter 1944–45 Battle of the Bulge. Note that the driver and co-driver have their hatches open, with the person on the viewer's right smoking a cigarette.

HALF-TRACKS AND ARMOURED CARS

M2 Half-track

During the late 1930s, the Ordnance Department evaluated the concept of half-tracked artillery tractors using French Citroën-Kégresse vehicles. In 1939, the White Motor Company (WMC) married the rear-located tracked running gear of the T9 half-tracked truck to the M3 Scout Car's front wheel arrangement.

Field testing with the ensuing prototype revealed that it was underpowered. Further developmental work unfolded, including the fitting of an up-rated engine. In late 1940, the US Army accepted this design for mass production by Autocar, Diamond T. Motor Car and WMC under the designation Half-track Car, M2.

In terms of its physical appearance, the front nose section of the M2 featured a single large rubber-tyred road wheel pairing. The remaining two-thirds of the hull, meanwhile, featured tracked running gear. The vehicle had a total length of 5.96m (19ft 7in), including the nose-mounted anti-ditching roller (or instead a Tulsa 18G winch). The track system comprised the following elements: a large closed-spoke 18-tooth frontal drive-sprocket wheel, a large open-spoked fixed rear idler wheel, one central VVSS bogie structure that split at the bottom into two horizontal arms that each supported two pairs of small solid road wheels and a single upper track return roller that sat on top of the bogie structure. The track's contact-with-ground length was 1.18m (3ft 11in). The front of the vehicle resembled a standard military truck with an engine cowling and driver's cabin. Behind this, stretching across the length of the tracks was a low-sided box-like open troop-carrying compartment. This could carry up to seven soldiers; there were two rows of three inward-facing seats and a rear-facing seat just behind the crew seating area.

M2 Half-track
Crew: 3 (+7 troops)
Production: 1941
Weight: 8.7 tonnes (8.6 tons)
Dimensions: (L,W,H): 5.96m (19ft 7in) x 1.96m (6ft 5in) x 2.3m (7ft 7in)
Engine: 110kW (147hp) White 160AX petrol (gasoline)
Road speed: 72km/h (45mph)
Range: 322km (200 miles)
Armament: 1 x 7.62mm (0.3in) M1919A4 MG
Armour: 6.4–12.7mm (0.3–0.5in)

M2 Half-track
This view of an M2 half-track nicely depicts the design's unusual rear track arrangement, with no fewer than four pairs of small road wheels derived from a single bogie array.

M2 Half-track

This M2, seen in Tunisia in January 1943, has a relatively unusual camouflage scheme of khaki 'clouds' daubed over the base olive drab finish.

Top speed

The M2 weighed 8.7 tonnes (8.6 tons) and was powered by a front-located 110kW (147hp) White 160AX 7-cylinder, four-cycle inline petrol (gasoline) engine. In terms of its transmission, it sported a five-gear Spicer 3461 system. The vehicle's running gear and power plant enabled it to obtain an impressive top speed of 72km/h (45mph) on tarmacked roads. The M2 had two fuel tanks located inside the rear sides of the troop compartment. Together, these carried 230 litres (51gal) of fuel. This enabled the M2 to obtain a maximum operational range on a single fuel load of 322km (200 miles).

In terms of firepower, the M2 half-track featured a lateral skate rail that ran across the vehicle's width located above the driving compartment. Anywhere on this rail could be attached a 12.7mm (0.5in) Browning M2HB heavy machine gun. An auxiliary weapon, namely the 7.62mm (0.3in) Browning M1919A4 machine gun, could also be attached to this rail. The vehicle carried 700 M2HB bullets and 7750 7.62mm (0.3in) rounds. The vehicle's ammunition lockers were located inside the front sides of the troop compartment, just behind the crew stations. The upper shelves were accessible from inside the vehicle, while the lower ones were accessed via small flaps in the outer hull armour plates. The M2 merely possessed limited levels of protection. The vehicle's bolted vertical hull side and rear plates were just 6.4mm (0.25in)

A column of nearly-completed M2 half-tracks awaits their final touches at the Diebold Safe & Lock Company production line, located in Canton, Ohio, December 1941.

HALF-TRACKS AND ARMOURED CARS

thick. Its frontal hull armour was between 6.4 and 12.7mm (0.3–0.5in) thick, sloped at 25 degrees from the vertical. The windshield cover had the thickest protection.

M2A1

From October 1943 onwards, Autocar and WMC transferred their manufacturing efforts to the improved M2A1 half-track car. In total, these firms produced 1643 M2A1s and converted a further 1266 M2s in the process of manufacture to the M2A1 configuration. At 8.9 tonnes (8.8 tons) the M2A1 weighed slightly more than its M2 predecessor.

The main difference between the two was that the M2A1 did away with the skate rail. Instead, the 12.7mm (0.5in) Browning M2HB heavy machine gun was fixed to its own substantial M49 ring mount, with a low shield-rail, situated above the co-driver's right-side seat. Three small tubular pintle mountings located on the troop compartment sides could sport the vehicle's second weapon, a 7.62mm (0.3in) Browning M1919A4 machine gun. The M2 and M2A1 provided sterling service as troop carriers to motorized infantry squads, as reconnaissance vehicles and as artillery-towing vehicles.

Here on 14 February 1943, an M4 Sherman from the US 1st Armored Division tows a broken-down M2 half-track near Sidi Bou Zid, Tunisia, during Operation 'Torch'.

M2A1 Half-track

Crew: 3 (+7 troops)
Production: 1943–44
Weight: 8.9 tonnes (8.8 tons)
Dimensions (L,W,H): 5.96m (19ft 7in) x 1.96m (6ft 5in) x 2.3m (7ft 7in)
Engine: 110kW (147hp) White 160AX petrol (gasoline)
Road speed: 69km/h (43mph)
Range: 322km (200 miles)
Armament: 1 x 12.7mm (0.5in) M2HB HMG; 2 x 7.62mm (0.3in) M1919A4 MG
Armour: 6.4–12.7mm (0.3–0.5in)

M2A1 Half-track

This M2A1 variant is distinguishable from its M2 predecessor by its different firepower arrangements; here, its M2HB heavy machine gun sits in an M49 ring mount, while two 7.62mm (0.3in) machine guns are fixed to two of the vehicle's three troop compartment side-located pintle mountings.

M3 Half-track

The M3 half-track was officially designated as the Carrier, Personnel, Half-track, M3. In essence, the design was an elongated M2 with 25cm (10in) added to the length of the troop-compartment.

Commencing May 1941, Autocar, Diamond T. and WMC manufactured a total of 12,391 M3s. Similar to the M2 half-track, the front of the M3 variant featured a single large front rubber-tyred road wheel pairing, while the remaining two-thirds of the hull sported tracks. The M3's VVS Suspension and running gear were largely similar to that of the M2, and also featured a large open-spoked rear idler wheel that had a double horizontal spring coil.

The M3 also featured a lengthened and redesigned troop-carrying compartment. Its capacity was up to 10 troops, with these being seated in two inward-facing rows of five seats. The compartment's rear had the useful addition of a door to facilitate the embarkation/disembarkation of the soldiers. A canvas canopy could be erected to cover this compartment.

Minimal protection

The M3 weighed 9.1 tonnes (9 tons), some 0.4 tonnes (0.4 tons) more than the M2. This vehicle was powered by the same 110kW (147hp) White 160AX petrol (gasoline) engine and sported the same Spicer 3461 transmission. The vehicle could achieve a creditable maximum by-road speed of 72km/h (45mph). The half-track carried 230 litres (51gal) in its fuel tanks, enabling it to obtain a maximum operational range on one fuel load of 322km (200 miles).

The vehicle merely enjoyed a limited degree of protection. Its frontal face-hardened steel hull armour was between 6.4 and 12.7mm (0.25–0.5in) thick sloped at 75 degrees from the horizontal. The vertical bolted hull side and rear plates were all just 6.4mm (0.25in) thick.

M3 Half-track

This M3 half-track has its canvas canopy erected to protect the troops from inclement weather; note also the anti-ditching roller sticking out from the the vehicle's hull nose.

M3 Half-track

Crew: 3 (+10 troops)
Production: 1941–43
Weight: 9.1 tonnes (9 tons)
Dimensions: (L,W,H): 6.16m (20ft 3in) x 1.96m (6ft 5in) x 2.3m (7ft 7in)
Engine: 110kW (147hp) White 160AX petrol (gasoline)
Road speed: 72km/h (45mph)
Range: 322km (200 miles)
Armament: 1 x 7.62mm (0.3in) M1919A4 MG
Armour: 6.4–12.7mm (0.25–0.5in)

HALF-TRACKS AND ARMOURED CARS

The only firepower the M3 half-track possessed was a solitary 7.62mm (0.3in) Browning M1919A4 machine gun. This was mounted with 360-degree traverse on a M25 pedestal mount affixed inside the front of the troop-carrying compartment.

M3A1

During October 1943 the first M3A1 half-tracks entered American Army service; this design was slightly heavier at 9.3 tonnes (9.2 tons). Up to March 1944, Autocar, Diamond T and WMC manufactured from scratch some 2862 M3A1 vehicles and converted 2209 extant or part-built M3s to the M3A1 specification. The main difference between the two designs was that the M3A1 dispensed with the M25 pedestal mount and instead featured an M49 or M49A1 Ring Mount. This housed – similar to the M2A1 half-track – a 12.7mm (0.5in) Browning M2HB heavy machine gun rather than the 7.62mm (0.3in) M1919A4 weapon seen in the M3. In addition, the M3/M3A1 served as the basis of several improvised half-tracked SPGs, including the T12 75mm (2.95in) Howitzer Motor Carriage.

Paratroopers from the US 17th Airborne Division, acting as infanteers, trudge along a snow-laden road next to an M3, near La Roche, Belgium, on 15 January 1945, during the Allied counter-offensive against the German Ardennes counter-strike.

M3A1 Half-track
Crew: 3 (+10 troops)
Production: 1941–43
Weight: 9.3 tonnes (9.2 tons)
Dimensions: (L,W,H): 6.16m (20ft 3in) x 1.96m (6ft 5in) x 2.3m (7ft 7in)
Engine: 110kW (147hp) White 160AX petrol (gasoline)
Road speed: 72km/h (45mph)
Range: 322km (200 miles)
Armament: 1 x 12.7mm (0.5in) M2HB HMG
Armour: 6.4–12.7mm (0.25–0.5in)

M3A1

This M3A1 is principally distinguished from its M3 predecessor by the large ring mount fixed to the top of the driving cab on which is mounted an M2HB heavy machine gun.

M5 Half-track

During much of 1942, the rate of manufacture of the M3/M3A1 half-track failed to keep up with the surging demand for newly-produced vehicles that emanated from both the American armed forces and, via Lend-Lease, from its principal Allies.

To remedy this, in December 1942 the Ordnance Department contracted IHC to manufacture a modified variant of the M3, designated the M5 half-track. In total, IHC constructed 4625 examples of the M5. America sent the majority of these vehicles to its Allies via Lend-Lease, notably to the United Kingdom.

Design differences

Given the different production methods and equipment at IHC, the M3 design had to be slightly modified. The two most significant differences were that the M5 was powered by an IHC-designed petrol (gasoline) engine and that its hull was made from welded RHA, often with rounded edges rather than the bolted face-hardened steel of the M2 and M3 with its right-angled edges. To bring the vehicle up to the same standard of protection as the M2 and M3, however, required slightly thicker plates. These changes

M5 Half-track

This M5, which like most of its sister vehicles was exported to an American ally, appears to bear the red-on-white triangular tactical symbol of the British 1st Infantry Division, seen here in Tunisia, spring 1943.

increased the vehicle's weight to 9.3 tonnes (9.2 tons); to compensate for this the M5 featured sturdier axles and hull strengthening modifications. The vehicle could carry, in addition to its three-person crew, up to 10 soldiers that could enter and exit the troop compartment through the door located in the vehicle's centre-rear.

Like the M3 – but unlike the M3A1 – the M5's firepower consisted of a solitary 7.62mm (0.3in) Browning M1919A4 machine gun. This weapon, with all-round traverse, was fitted on an M25 pedestal mount affixed inside the front of the vehicle's troop compartment. The vehicle carried

M5 Half-track
Crew: 3 (+10 troops)
Production: 1943
Weight: 9.3 tonnes (9.2 tons)
Dimensions: (L,W,H): 6.33m (20ft 9in) x 2.21m (7ft 3in) x 2.74m (9ft)
Engine: 107kW (143hp) International Harvester RED-450-B petrol (gasoline)
Road speed: 68km/h (42mph)
Range: 322km (200 miles)
Armament: 1 x 7.62mm (0.3in) M1919A4 MG
Armour: 6.4–16mm (0.25–0.625in)

4000 rounds for this weapon in stowage lockers located inside the front sides of the troop compartment. In terms of armoured protection, the M5's frontal hull armour was 6.4–16mm (0.25–0.625in) thick, sloped at 23 or 27 degrees from the vertical. The half-track's vertical welded RHA hull side and rear plates, moreover, were 8mm (0.31in) thick.

For its power plant, the three-crewed M5 featured a 107kW (143hp) International Harvester RED-450-B, six-cylinder, four-cycle, in-line petrol (gasoline) engine, which developed 472 Joules (348ft-lbs) of torque at 800rpm. This engine was marginally less powerful than the White one found in the M2 and M3 designs. The RED-450-B power plant was married to a different five-gear transmission known as the Spicer 1856. The slight decrease in power and modest increase in weight (compared to the M3) led to a minor decrease in the half-track's mobility. Thus, instead of the 72km/h (45mph) of the M3, the M5 half-track could only obtain a top speed of 68km/h (42mph). This design carried the same maximum fuel load – 230 litres (51gal) – as the M2 and M3 vehicles; consequently, the M5 could obtain the same maximum operational range of one fuel load of 322km (200 miles).

Canvas screen

From October 1943, the improved IHC-manufactured M5A1 half-track entered American service; IHC ultimately produced 2959 examples of this variant over the next year. Many of these were exported to America's Allies in the war, including Britain, the Free French and the Soviet Union. In British service the M5A1 was employed as the standard tractor to tow the Ordnance QF 6-pounder (57mm, 2.2in) and 17-pounder (76mm, 3in) anti-tank guns.

The main difference with the M5A1 sub-variant was that it featured lethality upgrades. Similar to the M2A1 and M3A1 vehicles, the M5A1 half-track sported a 12.7mm (0.5in) Browning M2HB heavy machine gun fixed to its own substantial M49/M49A1 pulpit ring mount situated above the co-driver's right-side seat. In addition, the vehicle housed three small tubular pintle mounts that could house the vehicle's 7.62mm (0.3in) M1919A4 machine gun. One of these mounts was to the right of the rear door and there was one each near the centre of the troop compartment sides.

Some vehicles, moreover, were completed with a second M1919A4 weapon as standard. Similar to the M5, the M5A1 half-track carried a canvas screen that could be used to cover the troop-carrying compartment using the two thin rails that ran across the compartment's top.

An M3 Scout Car, vehicle 6013770, seen at the US Army's Oak Ridge training facility.

M3 Scout Car

The M3 Scout Car (or the White Scout Car in British Army parlance) was initially designated as the M2A1, a modified variant of the M2 scout car; some 22 of the latter were built during 1935–38. The M3's main tactical roles were reconnaissance and screening.

During 1938–39, WMC manufactured 100 M3s. This design was a compact protected scout car with a front and rear pair of driving road wheels shielded by large curving fenders. The centre of the vehicle sported a shallow raised driving position, while the rear half comprised the open-topped troop-carrying compartment, the sides of which sloped down to the vehicle's vertical hull rear face that housed the exit/entry door.

Operational range

The 4x4 M3 Scout Car weighed 3.69 tonnes (3.63 tons) and had a two-man crew consisting of driver and a commander; six troops could be carried in the rear. In terms of armament, the arrangement of the following machine guns varied.

Many M3s were armed with a 12.7mm (0.5in) Browning M2HB heavy machine gun and one or two 7.62mm (0.3in) Browning M1919A4 machine guns all fitted to a skate mount. Some vehicles, however, were finished with one or two of these weapons missing. The M3 was protected by bolted-together rolled face-hardened steel plates. On the windshield, this armour was 12.7mm (0.5in) thick, while all other surfaces were just 6.4mm (0.25in). The M3 was powered by the standard 71kW (95hp) Hercules JXD six-cylinder petrol (gasoline) engine, which permitted it to reach a maximum road speed of 89km/h (55mph). The vehicle featured a 100 litre (22gal) fuel load giving it a maximum operational range of 402km (250 miles).

During early 1939, WMC developed an improved and widened variant of the M3 designated the M3A1, which at 5.62 tonnes (5.5 tons) weighed significantly more than its predecessor. American firms subsequently mass-manufactured the M3A1 through to

M3 Scout Car

On this M3 scout car you can clearly see the bolts of the two plates that formed the sides of the rear-located troop-carrying compartment. This vehicle is armed with two machine guns – one heavy and one medium.

M3A1 Scout Car

Crew: 2 (+6 troops)
Production: 1939–44
Weight: 5.62 tonnes (5.5 tons)
Dimensions: (L,W,H): 5.62m (18ft 5in) x 2m (6ft 8in) x 1.99m (6ft 6in)
Engine: 82kW (110hp) Hercules JXD petrol (gasoline)
Road speed: 72km/h (45mph)
Range: 402km (250 miles)
Armament: 1 x 12.7mm (0.5in) M2HB HMG; 1 x 7.62mm (0.3in) M1919A4 MG
Armour: 6.4–12.7mm (0.25–0.5in)

HALF-TRACKS AND ARMOURED CARS

summer 1944, by which time 20,918 examples had been completed. Thousands of these vehicles were dispatched to America's Allies.

The M3A1 featured six other main new features. Firstly, it was powered by the up-rated 82kW (110hp) Hercules JXD petrol (gasoline) engine. This mitigated against the vehicle's increased weight to a degree, but the M3A1's top road speed nevertheless still declined to 72km/h (45mph). Secondly, the M3A1 carried an anti-ditching roller device on the hull nose. Third, the new variant sported three frame rails over which a canvas screen could be erected. Fourth, its fuel load capacity was increased by 10 per cent. Fifth, its machine-gun skate rail was lowered slightly. Finally, the four-wheel drive mechanism was rendered permanent and could not be disengaged as in the M3.

Firms also manufactured 3340 M3A1E1 sub-variants for the Soviet Union, which had a Buda-Lanova 6DT-317 six-cylinder diesel engine.

M3A1 Scout Car

This M3A1, vehicle 'T17', serving with the Free French forces, has a lowered skate rail for its 12.7mm (0.5in) Browning M2HB heavy machine gun.

M3 Scout Car

This US Army M3 scout car, sporting a three-tone camouflage scheme, has been fitted with armoured plates to the driving compartment's side window-areas.

M1 Armoured Car

During 1931 the Ordnance Department directed the James Cunningham and Son Company of Rochester, New York, to develop the T4 armoured car prototype. Cunningham manufactured two prototype vehicles, while RAI produced a further four; all had some very minor design modifications as different elements were tested.

M1 Armoured Car

An M1 armoured car bears the distinctive crossed-swords tactical insignia of the US 1st Cavalry Regiment, 1938; note the unusual topped-off conical shape of its rear-located turret.

The 1st Cavalry Regiment trialled the use of these vehicles during 1932–33 at Fort Riley, Kansas. At this time the US Cavalry remained wedded to using horses for tactical mobility rather than armoured vehicles. After further development work, the Army accepted the design for a limited production run during 1934 as the M1. In total, Cunningham and RAI completed 20 T4/M1 armoured cars. The 1st Regiment continued to use these vehicles on an experimental basis until 1939 when they were withdrawn into reserve status. The experience gained with the M1 helped facilitate the subsequent development of the M18 Greyhound armoured car.

Classic shape

In terms of physical appearance, the front hull of the M1 featured a long angular engine housing with a single pair of front-located non-driving but steering rubber-tyred road wheels. Just behind and above these wheels the vehicle sported a spare pair of replacement wheels. The hull then rose sharply up at the driver's station with a near-vertical windshield.

The flat hull roof soon took a short step down until the upper rear of the vehicle, with a steeply-plunging rear hull face. The rear half of the chassis sat on two large driver road wheels, hence the designation 4x6. On top of the rear hull roof was a shallow conical turret with inward-sloping upper faces. Most of its small circular roof comprised the commander's flat hinged hatch.

Another distinctive feature was the vehicle's two large fenders that protected the top of the front and rear wheels; the front fenders each housed a large forward-directed head-lamp.

M1 Armoured Car
Crew: 4
Production: 1931–34
Weight: 4.2 tonnes (4.1 tons)
Dimensions: (L,W,H): 4.57m (15ft) x 1.83m (6ft) x 2.1m (6ft 11in)
Engine: 99kW (133hp) Cunningham petrol (gasoline)
Road speed: 89km/h (55mph)
Range: 402km (250 miles)
Armament: 1 x 12.7mm (0.5in) M2HB HMG; 1–3 x 7.62mm (0.3in) M1919A4 MG
Armour: 6.4–9.5mm (0.25–0.375in)

Armament

In terms of firepower, the M1's turret front carried a 12.7mm (0.5in) Browning M2HB heavy machine gun. A few vehicles also featured a roof-top bracket to house a 7.62mm (0.3in) M1919A4 machine gun. Many turrets also featured a fore and aft pistol port, through which two more M1919A4 machine guns could be operated. The M1 was only lightly protected with 6.4–9.5mm (0.25–0.375in) thick welded and bolted armoured plates. The M1 weighed 4.2 tonnes (4.1 tons) and was manned by a crew of four: the commander, driver, gunner and loader.

The vehicle was powered by a front-located 99kW (133hp) Cunningham V8-cylinder, four-cycle, liquid-cooled petrol (gasoline) engine. With its four rear-driver and two front idler wheels, the M1 could obtain an impressive maximum speed along tarmacked roads of 89km/h (55mph). On a single fuel load of 114 litres (25gal) the M1 could obtain a creditable maximum operational range of 402km (250 miles).

M8 Armoured Car

During 1941, the Ordnance Department requested tenders for a new armoured car design to be employed mainly in the anti-tank role. The project's specifications included a highly-mobile 6x6 vehicle with sloped armour, a turret-mounted 37mm (1.46in) M6 cannon as main armament and several secondary weapons.

An M8 from the US 30th Division, with its crew in exposed positions, moves down a track in the heavily damaged village of Kinzweiler, north-east of Aachen, Germany. Note the ditched and abandoned German assault gun in the background.

HALF-TRACKS AND ARMOURED CARS

M8 Armoured Car
This late-production M8 Greyhound, seen here during the 1944 Normandy campaign, features a prominent large ring mount for its M2HB heavy machine gun.

In response, the T21, T22 and T23 prototypes were respectively developed by Studebaker, Ford and Chrysler. After extensive testing during the winter of 1941–42, the Army selected the Ford T22E2 design for mass production.

However, soon thereafter it was recognized that the 37mm (1.46in) M6 gun was now inadequate as a tank-killer against the latest generation of Axis AFVs. Consequently, rather than cancel the production run, the Ordnance Department re-defined the T22E2 as the Light Armoured Car, M8. Early manufacturing problems stalled production, which only got underway in March 1943. Thereafter, American firms produced 8523 Greyhounds, as they became known, by the war's August 1945 end.

Weighing 7.9 tonnes (7.8 tons), the M8 was manned by a crew of four: the commander (turret right), the gunner (left turret), the driver (hull front left) and the radio operator/co-driver in the hull front right. The M8 was a compact vehicle with a shallow mostly-welded hull, upon which sat a tall and large centrally-located open-roofed two-man turret.

The 6x6 vehicle had one front pair of driving wheels and two driving rear pairs; steering was via the front wheels only. The Greyhound was powered by the up-rated 82kW (110hp) Hercules JXD six-cylinder petrol (gasoline) engine. This enabled the vehicle to reach a top speed of 90km/h (56mph) on tarmacked roads, although its cross-country performance never reached the levels expected of the design. On a single load of fuel, the M8 could obtain a maximum cruising operational range of 563km (350miles).

Turret armour
The front of the turret sported a 37mm (1.46in) M6 cannon aimed via a M70D telescopic sight, for which the vehicle carried up to 80 rounds. The turret also featured a co-axial 7.62mm (0.3 in) M1919A4 machine gun, for which 1500 rounds were carried. Early production vehicles also sported a pintle mount on the hull roof behind the turret for

M8 Armoured Car
Crew: 4
Production: 1943–45
Weight: 7.9 tonnes (7.8 tons)
Dimensions: (L,W,H): 5m (16ft 5in) x 2.5m (8ft 2in) x 2m (6ft 8in)
Engine: 82kW (110hp) Hercules JXD petrol (gasoline)
Road speed: 90km/h (56mph)
Range: 563km (350 miles)
Armament: 37mm (1.46in) M6; 1 x 12.7mm (0.5in) M2HB HMG; 1 x 7.62mm (0.3in) M1919A4 MG
Armour: 9.5–25mm (0.375–1in)

an externally-operated 12.7mm (0.5in) M2HB Browning heavy machine gun. Late-production examples, however, instead featured a large ring mount over the turret roof from which the M2HB Browning could be operated. To save weight and maintain mobility, the vehicle was only very lightly protected with thin welded steel RHA plates. The M8 sported 9.5–19mm (0.375–0.75in) thick frontal hull plates sloped at 30–60 degrees from the vertical, as well as 9.5mm (0.375in) side and rear hull plates. Its turret had armour that varied in thickness from 6.4mm (0.25in) to 25mm (1in). The vehicle also sported a powerful SCR506, 510, 608 or 610 wireless-radio.

On top of the rear of this M8, which has lost its side skirts, three US Army soldiers perch, weapons ready in case of a contact with the enemy, somewhere in Northwest Europe during the winter of 1944–45.

HALF-TRACKS AND ARMOURED CARS

M8 Armoured Car
This M8 has been captured and pressed into German service, with German national insignia on the turret side; the swastika flag draped over the turret rear might be an aerial recognition measure to help prevent attack by friendly Luftwaffe aircraft.

M8 Armoured Car
This early-production M8, seen in Sicily, 1943, is identified as such by the tall mounting for its external Browning M2HB machine gun located on the hull roof just behind the turret rear.

HALF-TRACKED SELF-PROPELLED GUNS

During the early and middle phases of World War II, the US Army developed a series of half-tracked self-propelled guns (SPGs) and self-propelled anti-aircraft guns (SPAAGs), based on the chassis of existing designs. These typically improvised mobile direct and indirect fire assets temporarily provided useful combat service until more effective, fully-tracked versions emerged in the war's later stages.

The following AFVs are featured in this chapter:
- 75mm Howitzer Motor Carriage T12
- 75mm Howitzer Motor Carriage T30
- 81mm Mortar Carrier M4
- Multiple Gun Motor Carriage M13
- Multiple Gun Motor Carriage M14
- T28E1 Combination Gun Motor Carriage
- Combination Gun Motor Carriage M15
- Multiple Gun Motor Carriage M16
- Multiple Gun Motor Carriage M17

After their forces captured intact the Rhine bridge at Remagen, Germany on 7 March 1945, American SPAAGs – like this MGMC M16 with its quad 12.7mm (0.5in) machine guns – assembled to prevent Luftwaffe attempts to destroy it.

HALF-TRACKED SELF-PROPELLED GUNS

75mm Howitzer Motor Carriage T12

The T12 was an early US Army half-tracked self-propelled gun (SPG)/improvised tank-destroyer. Based on the M3 half-track chassis, the T12 featured a 75mm (2.95in) M1897A4 gun fixed above and behind the driver's cab position in a T30 mounting with a small gun-shield. The gun had a limited arc, with a traverse to the left or right of only 22 degrees; to engage targets beyond this traverse the vehicle itself would have to be moved.

This four-crew SPG carried a total of 59 rounds for its main armament. The T12 weighed 9.1 tonnes (9 tons) and was powered by a 110kW (147hp) White 160AX petrol (gasoline) engine. The vehicle also sported a Browning M1919A4 machine gun fixed to the back of the superstructure. Its armoured protection merely ranged from between 6.4mm (0.25in) and 12.7mm (0.5in) thick bolted plates; this provided inadequate crew survivability during the tactical mission of engaging enemy tanks.

Indirect fire-support

WMC produced 2203 T12s, many of which served in the North African and Italian campaigns. By 1944, the vehicle's anti-armour capabilities were inadequate against the newer, better-protected German Panzer IV (Models H-M), V and VI tanks; it was thus increasingly used in an indirect fire-support role. The T12 proved a better tank-killer in the Pacific theatre against the less well-protected Japanese AFVs and delivered effective service during the 1943–45 battles of Tarawa, Saipan and Okinawa.

T12 HMC

Two features help differentiate the T12 from its sister improvised M3 half-track SPG conversion, the T30; the T12's 75mm howitzer had a much longer barrel and its small gun shield was of a distinctive three-sided wing-like design.

75mm Howitzer Motor Carriage T12
Crew: 4
Production: 1941–42
Weight: 9.1 tonnes (9 tons)
Dimensions: (L,W,H): 6.16m (20ft 3in) x 1.96m (6ft 5in) x 2.3m (7ft 7in)
Engine: 110kW (147hp) White 160AX petrol (gasoline)
Road speed: 68km/h (42mph)
Range: 322km (200 miles)
Armament: 75mm (2.95in) M1897A4; 1 x 7.62mm (0.3in) M1919A4 MG
Armour: 6.4–12.7mm (0.25–0.5in)

75mm Howitzer Motor Carriage T30

The T30 HMC was another early American half-tracked self-propelled gun (SPG). The vehicle mounted upon a modified M3 half-track chassis a M1A1 75mm (2.95in) L/18.4 Pack Howitzer. The first prototype vehicle was developed during late 1941.

The M1A1 gun featured a large three-sided, frontally steeply angled gun-shield and sat upon a box structure set inside the rear-located troop-carrying superstructure. This Howitzer could fire (indirectly) rounds out to a maximum range of 8800m (9624yds). The T30 also sported a rear-located pedestal-mounted M2HB Browning heavy machine gun. Weighing 9.3 tonnes (9.2 tons), the T30 was powered by a 110kW (147hp) White 160AX petrol (gasoline) engine.

Limited survivability

During February–April 1942, WMC delivered 314 T12 SPGs to the Army, with a final batch of 188 in November. With its limited survivability and modest anti-armour firepower, the T30 was obviously a stop-gap expedient that temporarily met a tactical requirement while more effective alternate fully-tracked designs were developed. The first examples of the T30 to enter combat did so during the November 1942 Operation Torch landings in Axis-controlled North Africa. Other vehicles subsequently served in both the Sicilian and Italian campaigns.

In combat

Combat experience, however, soon showed that these improvised SPGs had little value in direct action against German tanks; consequently, they were used as indirect fire-support weapons. Most T30s were subsequently replaced by the fully-tracked M8 HMC based on the M5 Stuart light tank chassis.

T30 HMC

In comparison with its sister improvised half-track SPG, the T12, the expedient T30 design had a different, shorter-barrelled 75mm gun and a larger three-sided triangular-profiled gun shield.

75mm Howitzer Motor Carriage T30
Crew: 5
Production: 194–4
Weight: 9.3 tonnes (9.2 tons)
Dimensions: (L,W,H): 6.1m (21ft) x 1.96m (6ft 5in) x 2.3m (7ft 7in)
Engine: 110kW (147hp) White 160AX petrol (gasoline)
Road speed: 72km/h (45mph)
Range: 322km (200 miles)
Armament: 75mm (2.95in) M1A1; 1 x 12.7mm (0.5in) M2HB HMG
Armour: 6.4–12.7mm (0.25–0.5in)

HALF-TRACKED SELF-PROPELLED GUNS

81mm Mortar Carrier M4

The six-crew M4 was another early-war American self-propelled indirect fire vehicle, but it sported an 81mm (3.2in) M1 mortar. This weapon was mounted facing backwards in the rear of the small M2 half-track's rear-located troop compartment.

The main tactical constraint of this weapon was that it was manually traversable only through a limited arc when fired from inside the vehicle, which was the preferred option. The 8-tonne (7.9-ton) M4 also sported a 7.62mm (0.3in) M1919A4 machine gun on an M35/M35C mobile mount that could be clamped to multiple positions.

Engine
Powered by the up-rated 110kW (147hp) six-cylinder White 160AX engine petrol (gasoline) engine, the M4 could achieve a maximum speed of 72km/h (45mph) along tarmacked surfaces.

The vehicle's bolted rolled face-hardened armoured plates typically were 6mm (0.25in) thick, with the windscreen cover being 12.7mm (0.5in) thick. During 1941–43, WMC manufactured 572 M4 mortar carriers.

M4A1 modification
From May 1943 onwards, WMC switched production from the M4 to the M4A1. The modified 8.2–tonne (8.1–ton) M4A1 carrier had its mortar married to an improved base-mount. This enabled the weapon to be manually traversed through a greater range.

This variant could in part be visually differentiated from its predecessor because the mortar stood up slightly higher beyond the top of the open troop compartment's edge. During 1943–44 WMC built a total of 600 M4A1s.

M4 Mortar Carrier
This M4A1 mortar carrier can be distinguished from its M4 predecessor because the mortar barrel projects to a greater degree above the vehicle's hull sides; the default position was to have the mortar facing backwards.

81mm Mortar Carrier M4
Crew: 6
Production: 1941–43
Weight: 8 tonnes (7.9 tons)
Dimensions: (L,W,H): 6.01m (19ft 9in) x 1.96m (6ft 5in) x 2.27m (7ft 5in)
Engine: 110kW (147hp) White 160AX petrol (gasoline)
Road speed: 72km/h (45mph)
Range: 322km (200 miles)
Armament: 1 x 81mm (3.2in) M1 mortar; 1 x 7.62mm (0.3in) M1919A4 MG
Armour: 6.4–12.7mm (0.25–0.5in)

Multiple Gun Motor Carriage M13

The Multiple Gun Motor Carriage (MGMC) M13 was a half-tracked self-propelled anti-aircraft gun vehicle (SPAAG) based on the M3 half-track chassis. During 1942, the US Army realized it urgently needed to protect its armoured columns from enemy aerial attack, and the M13 was an early response to this requirement.

During July 1942–May 1943, WMC produced 1103 M13 vehicles; however, half of these were converted to the M16 configuration before reaching the US field army.

Only 139 M13 vehicles reached the US Army forces waging World War II in Europe. Most of these M13s served with the Fifth US Army during its early 1944 amphibious landings at Anzio, Italy. By March 1944, new arrivals of the MGMC M16 vehicle began to replace the surviving M13s in Italy.

Modified M3 half-track

The five-crew M13 was a modified M3 half-track that sported at its rear an M33 Maxson open-topped mounting that held a pair of remotely-controlled 12.7mm (0.5in) Browning M2-TT-HB heavy machine guns, each of which could deliver 570rpm.

The 8.2-tonne (8-ton) M13 was powered by a 110kW (147hp) White 160AX petrol (gasoline) engine and could achieve a top speed of 72km/h (45mph). Its 230-litre (51-gal) fuel capacity enabled the M13 to obtain a maximum operational range of 282km (175 miles).

The M13 briefly gave useful service in Italy but was soon superseded by the MGMC M16's superior firepower.

M13 MGMC

This SPAAG's crew has dubbed this MGMC M13 'Nasty Lucy', along with some racy figurine art; it mounts a twinned pair of 12.7mm (0.5in) Browning M2-TT-HB heavy machine guns in an M33 Maxson mounting.

Multiple Gun Motor Carriage M13
Crew: 5
Production: 1942–43
Weight: 8.4 tonnes (8.3 tons)
Dimensions: (L,W,H): 6.5m (21ft 4in) x 1.98m (6ft 6in) x 2.2m (6ft 8in)
Engine: 110kW (147hp) White 160AX petrol (gasoline)
Road speed: 72km/h (45mph)
Range: 282km (175 miles)
Armament: twinned x 12.7mm (0.5in) M2-TT-HB HMG
Armour: 6.4–12.7mm (0.25–0.5in)

HALF-TRACKED SELF-PROPELLED GUNS

Multiple Gun Motor Carriage M14

The Multiple Gun Motor Carriage (MGMC) M14 was another mid-war SPAAG intended to suppress low-flying enemy ground-attack aircraft.

Unlike its sister design, the M13, the M14 was based on the chassis of the M5 half-track. The M14 weighed more than its sister – at 9.4 tonnes (9.25 tons) – due to the M5 half-track's larger chassis. The M14 design featured the same rear-mounted M33 Maxson open-topped mounting with two 12.7mm (0.5in) Browning M2-TT-HB heavy machine guns fitted as its sister M13 vehicle.

The M14 vehicle had modest armour protection of between 6mm (0.24in) and 16mm (0.625in) thickness. Its 107kW (143hp) IH RED-450-B petrol (gasoline) engine enabled the vehicle to reach a maximum on-road speed of 68km/h (42mph).

UK export

From mid-1942 onwards IHC began producing M14 SPAAGs intended for export to the United Kingdom under

M14 MGMC

This M14 SPAAG in American service has been fitted with a powerful winch adjacent to its frontal anti-ditching roller; it carries the same M33 Maxson weapon system as its predecessor M13.

the Lend-Lease Act. The original order was for 1600 vehicles. By the time that just 200 had been built, however, the Americans had developed the far superior M45 Quad-mount, which sported four 12.7mm (0.5in) Browning M2 heavy machine guns.

The Americans increasingly diverted M13 and M14 chassis to be fitted with the M45 Quad-mount, thus creating the M15 and M16 SPAAGs, respectively based on the M3 and M5 half-track chassis. Of the few M14s to reach the British Army, most had their weapons removed during conversion to carrier vehicles.

Multiple Gun Motor Carriage M14

Crew: 5
Production: 1943
Weight: 9.4 tonnes (9.3 tons)
Dimensions: (L,W,H): 6.49m (21ft 4in) x 2.18m (7ft 2in) x 2.3m (7ft 7in)
Engine: 107kW (143hp) International Harvester RED-450-B petrol (gasoline)
Road speed: 68km/h (42mph)
Range: 322km (200 miles)
Armament: twinned x 12.7mm (0.5in) M2-TT-HB HMG
Armour: 6.4–16mm (0.25–0.625in)

T28E1 Combination Gun Motor Carriage

The T28E1 was the developmental 'father' of the M15 SPAAG. The design derived from the earlier T28 project. The latter design was developed for the US Army's Coastal Artillery units.

As a development of the T1A2 MGMC project, the T28 CGMC mounted a 37mm (1.46in) gun on a M2 half-track. During 1942 tests at the Aberdeen Proving Ground revealed that the 37mm (1.46in) weapon suffered from excessive recoil, and this rendered the light M2 half-track platform unstable. The project was thus subsequently suspended.

Operation Torch

Next, the US Army began its preparations for Operation Torch, the November 1942 Western Allied invasion of Axis-held North Africa. During these preparations there resurfaced the requirement for an improvized SPAAG that mounted a 37mm (1.46in) cannon. This led to the T28 project being re-started, with this work morphing into the development of the T28E1 CGMC.

High vulnerability

The T28E1 SPAAG married to the M3 half-track chassis an open mount that sported a single 37mm (1.46in) M1 L/54 cannon alongside pair of 12.7mm (0.5in) Browning M2HB heavy machine guns; hence the designation 'Combination' rather than 'Multiple' weapons. The mount, which was the predecessor of the M42 mounting, lacked any gun-shield which rendered both the crew and gun highly vulnerable to enemy fire.

During July and August 1942 WMC produced a total of 80 T28E1s. Meanwhile, further development work ensued which led to the creation of the CGMC M15 during late 1942.

T28E1 Combination Gun Motor Carriage
Crew: 7
Production: 1942
Weight: 9.1 tonnes (9 tons)
Dimensions (L,W,H): 6.17m (20ft 3in) x 2.24m (7ft 4in) x 2.39m (7ft 10in)
Engine: 110kW (147hp) White 160AX petrol (gasoline)
Road speed: 68km/h (42mph)
Range: 241km (150 miles)
Armament: 37mm (1.46in) M1A2; 2 x 12.7mm (0.5in) M2HB HMG
Armour: 6.4–12.7mm (0.25–0.5in)

T28E1
The sheer lack of protection afforded to the seven-man crew by this early-war T28E1 improvised SPAAG is all too apparent in this side view.

HALF-TRACKED SELF-PROPELLED GUNS

Combination Gun Motor Carriage M15

The M15 was the mid-war American SPAAG developed from the T28E1. The M15 mounted a 'Combination' of one cannon and two machine guns in a rear-located M42 armoured mounting. This sported a 37mm (1.46in) M1A2 L/54 cannon, above which sat a pair of 12.7mm (0.5in) Browning M2HB heavy machine guns.

The vehicle, based on the M3 half-track chassis, sported thin 6.4–12.7mm (0.25–0.5 in) armoured plates. WMC and Autocar produced 682 M15s during 1943. The M15 was powered by the same 110kW (147hp) White 160AX petrol (gasoline) engine as its sister SPAAGs. The M15 design similarly featured 230-litre (51-gal) fuel tanks and could obtain 68km/h (42mph). The vehicle saw service in Operation Torch, the Allied invasion of Axis-held North Africa.

Aircraft killer

While these combat experiences showed that the M15 was a potent aircraft-killer, it also revealed that the M45 mount placed stress on the chassis, causing maintenance problems. This led to the development, by late 1943, of the modified M15A1 variant. This sported the improved M54 mounting, which placed the Browning machine guns below, rather than above, the M1 cannon.

The marrying of the M54 mount on a modified M3A1 chassis created the M15AI variant. WMC and Autocar produced 1053 M15A1s during 1943–44. These vehicles served in the anti-aircraft battalions found in US armoured divisions, and they provided great suppressive effect in both the European and Pacific theatres.

M15 CGMC

In this M15 (rather than M15A1) SPAAG, the twin Browning machine guns were located above the 37mm cannon, all set inside a large three-sided box-like crew shield.

Combination Gun Motor Carriage M15
Crew: 7
Production: 1943
Weight: 9 tonnes (8.9 tons)
Dimensions: (L,W,H): 6m (19ft 4in) x 2.5m (8ft 2in) x 2.64m (8ft 8in)
Engine: 110kW (147hp) White 160AX petrol (gasoline)
Road speed: 68km/h (42mph)
Range: 322km (200 miles)
Armament: 37mm (1.46in) M1A2; 2 x 12.7mm (0.5in) M2HB HMG
Armour: 6.4–12.7mm (0.25–0.5in)

Multiple Gun Motor Carriage M16

During 1942, as M13 and M14 production unfolded the Army also developed a superior anti-aircraft weapon, the Maxson M45 Quad-mount: a modified M33 Maxson mount that sported four Browning M2HB machine guns.

M16 MGMC

The four Browning heavy machine guns mounted on the M16 SPAAG made it a significantly more effective design than any of its predecessors.

With each gun capable of 575rpm, for short periods the Quad-mount could fire an impressive 2300 12.7mm (0.5in) bullets into the air within a minute. Allied tactical doctrine at the time held that if you put enough metal into the air, some of it would ultimately hit an enemy aircraft.

Weighing 1.1 tonnes (1.08 ton), the M45 had a 360-degree traverse and an elevation range of between -5 and +90 degrees. When in February 1943 this mounting was married to a modified M3 half-track chassis (the T1E1), the five-crew MGMC M16 design was created.

Excellent suppressive capabilities

Weighing 9 tonnes (8.9 tons), the M16 was powered by the same machine as in the M13, the 110kW (147hp) White 160AX petrol (gasoline) engine.

With a reduced hp/weight ratio compared with its predecessor, this SPAAG could achieve a slightly less impressive top speed of 68km/h (42mph). Its 230-litre (510-gal) fuel capacity also enabled the MGMC M16 to obtain a maximum operational range of 282km (175 miles).

Production

The WMC produced 3550 M16s during 1942–44, which included 568 M13s converted to this more potent configuration. The M16 provided excellent suppressive capabilities to the American forces deployed in Europe and in the Pacific.

Multiple Gun Motor Carriage M16
Crew: 5
Production: 1944
Weight: 8.9 tonnes (8.8 tons)
Dimensions: (L,W,H): 6.5m (21ft 4in) x 1.98m (6ft 6in) x 2.2m (6ft 8in)
Engine: 110kW (147hp) White 160AX petrol (gasoline)
Road speed: 68km/h (42mph)
Range: 282km (175 miles)
Armament: 4 x 12.7mm (0.5in) M2-TT-HB HMG
Armour: 6–12.7mm (0.25–0.5in)

HALF-TRACKED SELF-PROPELLED GUNS

Multiple Gun Motor Carriage M17

The MGMC M17 was the sister of the M16, mounting the potent Maxson M45 12.7mm (0.5in) Browning M2HB machine gun Quad-mount on the larger M5 half-track chassis.

IHC commenced M17 production in December 1943 and over the course of an 11-month production run they produced 1002 vehicles. The United States Army dispatched the entire M17 production run to the Soviet Union under the Lend-Lease agreement. These formidable aircraft-killers provided excellent service to the Red Army during key offensives such as the summer 1944 Bagration and the early 1945 Vistula-Oder operations.

Fully-traversable mount
Weighing 8.9 tonnes (8.8 tons), the five-crew MGMC M17 was powered by the 107kW (143hp) IH RED-450-B petrol (gasoline) engine. This motive power-base enabled the vehicle to reach a maximum speed of 68km/h (42mph) along roads over a maximum operational range of 322km (200 miles). This SPAAG featured 6.4mm (0.25in) thick armoured plates, as well as a windscreen visor that was 12.7mm (0.5in) thick.

Weaponry
The vehicle used the modified Maxson M45D mounting of four M2HB Browning heavy machine guns, for which it carried 5000 rounds. This fully-traversable mount, fitted with a Reflex Sight M18 or an Illuminated Sight Mk. 9, had a slightly increased elevation range of between -10 and +90 degrees. Finally, the M17 sported a SCR528 wireless radio for external communication capability.

M17 MGMC

The M17, the IHC-produced version of the M16, could be distinguished from its sister design by several differences, most notably the lower height of the rear portion of the hull's sides.

Multiple Gun Motor Carriage M17
Crew: 5
Production: 1943–44
Weight: 8.9 tonnes (8.8 tons)
Dimensions: (L,W,H): 6.5m (21ft 4in) x 2.18m (7ft 2in) x 2.3m (7ft 7in)
Engine: 107kW (143hp) International Harvester RED-450-B petrol (gasoline)
Road speed: 68km/h (42mph)
Range: 322km (200 miles)
Armament: 4 x 12.7mm (0.5in) M2-TT-HB HMG
Armour: 6.4–12.7mm (0.25–0.5in)

Index

References to images are in *italics*.

A
AFVs (armoured fighting vehicles) 6–7
Ardennes, Battle for the 44, *48*, *104*

B
British Army 15, 16, 17
    1st Infantry Division *105*
    CDLs 39
    Lee/Grant medium tank 34–8
    Sherman medium tank 41, 51–2, *54*
Bulge, Battle of the *77*, *98*
Burma 35

C
Chaffee, M24 26–29
China 10, *49*
Combination Gun Motor Carriage (CGMC):
    M15 122
    T28E1 121

D
Dutch East Indies 10

E
El Alamein, Battle of 38, 51, 52

France 22
    Ardennes *44*, *104*
    M18 Hellcat *76*, *77*
    Normandy *21*, *73*
Free French Army *18*
    HMC M7 83
    M3 Scout Car *108*
    M10 74
    Sherman medium tank 52–3

G
German Army *113*
German tanks 68–9
Germany 27, *30*, *50*, *67*
    Aachen *110*
    Siegfried Line 57
GMC DUKW *97*
Grant, Gen Ulysses S. 34
Guadalcanal *8*

Gun Motor Carriage (GMC)
    M12 155mm 85
    M19 Multiple 7, 87
    M40 155mm 7, 86

H
Howitzer Motor Carriage (HMC)
    M7 Priest 105mm 7, 83
    M8 81–2
    M37 105mm 84
    T12 75mm 7, 116
    T30 75mm 117

Italy *45*, *70*, *74*
Iwo Jima 44, *46*, *88*
    LVTs *95*, *96*

J
Japan 10, 19, 116

K
Kwajalein Atoll *92*

L
Lee, Gen Robert E. 34
Lee/Grant M3 tank 34–38
Lend-Lease programme 10, 16

INDEX

Locust, M22 24–25
LVT (A)-1 93
LVT-1 Mk1 Alligator 7, 90–1
LVT-2 Water Buffalo 92
LVT-3 Bushmaster 7, 94
LVT-4 Water Buffalo 7, *88*, 95
LVT(A)-4 7, 96

M

M1 armoured car 109–10
M2:
    A4 light tank *8*, 12–14
    half-track APC 7, 100–2
    Lee/Grant medium tank 6–7, 32–3
M3:
    A1 Satan 19
    half-track APC 7, 103–4
    Lee Canal Defence Light (CDL) 39–40
    Lee/Grant medium tank 6–7, 34–8
    scout car 107–8
    Stuart light tank 6, 15–18
M4:
    81mm half-track Mortar Carrier 7, 118
    A3 Sherman 7, *30*, 41–54, *102*
    A3E2 Jumbo Assault tank 57
    A3R5 Flamethrower 7, 55
    M4A2 Flail Tank 56
    M4A4 Crab Mk. I 54
M5:
    half-track 105–6
    Stuart 20–3
M6 heavy tank 62–63
M6 A2E1 heavy tank 7, 62–3, 64
M7 Priest 58
M8 Greyhound armoured car *98*, 110–13
M10 Wolverine tank destroyer 7, *70*, 72–4
M15 half-tracked SPAAG 7
M18 Hellcat tank destroyer *6*, 7, 75–7
M22 Locust airborne light tank 7, 24–5
M24 Chaffee light tank 7, 22, 26–9
M26 Pershing heavy tank 7, *60*, 65–7
M26E4 'Super Pershing' 68–9
M36 tank destroyer 7, 78–80
Marmon-Herrington 24
    CTLM/CTLS 10
Massey Harris Company 20, 27
Multiple Gun Motor Carriage (MGMC):
    M13 7, 119
    M14 7, 120
    M16 *114*, 123
    M17 124

O

Okinawa 94, *96*, 116
operations:
    Plunder 25
    Torch 17, *20*, 36–*8*, 50
    Varsity 25

P

Pershing, John J. 66
Pershing, M26 heavy tank 65–69

INDEX

R

Rhine River *60*, *114*

Royal Dutch East Indies Army (RNIL) 10

S

Saipan 55, 116

Sherman tank 41–54

Sicily *113*

Stuart tank, M3 15–18

Stuart tank, M5 20-23

T

T1:
    'Cunningham' 11
    E3 'Aunt Jemima' 49–50

T23 59

Tarawa, Battle for 50, *91*, 116

Tunisia *36–8*, 50
    half-tracks *101*, *102*, *105*

U

US Army:
    1st Armored Dvn 29, *45*, *102*
    3rd Armored Dvn *21*
    9th Armored Dvn 27
    12th Armored Dvn *22*
    17th Airborne Dvn *104*
    30th Dvn *110*
    70th Infantry Dvn *30*
    1st Cavalry Rgt *109*

13th Armored Rgt *36*

66th Armored Rgt *23*

14th Tank Btn *60*

36th Tank Btn *50*

69th Tank Btn *44*, *48*

701st Tank Destroyer Btn *70*

740th Tank Btn *26*

US Marine Corps (USMC) 56, *92*

4th Dvn 56

10th Armored Dvn *42*

3rd Tank Btn *16*

4th Tank Btn *22*, *44*, *46*

Picture Credits

Artworks:
All artworks courtesy David Bocquelet/Tank Encyclopedia, except pages 46–47 and 54, courtesy Amber Books Ltd.

Photographs:
AirSeaLand.photos: 6, 8, 18, 22, 27, 33, 45, 50, 70, 73, 74, 76, 77, 82, 88, 92, 101, 102, 104, 110, 112, 114

Alamy: 13 (Prestor Pictures)

Amber Books: 7, 28, 40, 60, 67, 98

Getty Images: 21 (Galerie Bilderwelt), 30 (Horace Abrahams/Keystone/Hulton Archive), 79 (Photo12/UIG)

Naval History and Heritage Command: 91

Public Domain: 63, 106